WITHDRAWN

The Making of the 1996 Farm Act

THE MAKING OF THE 1996 FARM ACT

Lyle P. Schertz
Otto C. Doering III

Property of
CENTRAL COMMUNITY COLLEGE
Hastings Campus

Iowa State University Press / Ames

LYLE P. SCHERTZ, PHD, is a former economist and agency manager in the U.S. Department of Agriculture. He was a staff member of the U.S. Senate Committee on Agriculture, Nutrition, and Forestry, assisting in preparing for what eventually became the 1996 Farm Act. Dr. Schertz is the founding editor of *Choices, the Magazine of Food, Farm, and Resource Issues*, published by the American Agricultural Economics Association.

OTTO C. DOERING III, PHD, is professor of agricultural economics at Purdue University with extensive experience in Washington working on farm legislation, including several previous farm bills. In addition to his teaching and research, Dr. Doering is active in public service at the state and national level.

© 1999 Iowa State University Press
All rights reserved
Copyright is not claimed for appendixes 1, 2, and 3.

IOWA STATE UNIVERSITY PRESS
2121 South State Avenue, Ames, Iowa 50014

Orders:	1-800-862-6657
Office:	1-515-292-0140
Fax:	1-515-292-3348
Web site:	www.isupress.edu

Authorization to photocopy items for internal or personal use, or the internal or personal use of specific clients, is granted by Iowa State University Press, provided that the base fee of $.10 per copy is paid directly to the Copyright Clearance Center, 222 Rosewood Drive, Danvers, MA 01923. For those organizations that have been granted a photocopy license by CCC, a separate system of payments has been arranged. The fee code for users of the Transactional Reporting Service is 0-8138-2608-X/99 $.10.

∞ Printed on acid-free paper in the United States of America

First edition, 1999

Library of Congress Cataloging-in-Publication Data
Schertz, Lyle P.
    The making of the 1996 Farm Act / Lyle P. Schertz, Otto C. Doering III.—1st ed.
      p. cm.
    Includes bibliographical references and index.
    ISBN 0-8138-2608-X
    1. United States. 1996 Farm Act legislative history. 2. Agricultural laws and legislation—United States legislative history. I. Doering, Otto C. III (Otto Charles). II. Title.
KF1681.A315A1673   1999
343.73'076'0262—dc21   99-28326

The last digit is the print number: 9 8 7 6 5 4 3 2 1

Contents

Preface   vii

Introduction   ix

Acronyms   xi

Senators and Representatives   xiii

1	Five Early Developments: April 1994 to April 1995	3
2	Getting Serious About Commodity Legislation, May–August 1995	21
3	Searching for Consensus in the Senate	39
4	The Administration's Restrained Entry	45
5	The State of Play as Congress Recessed in August 1995	55
6	Counting Votes in the Fall of 1995	61
7	A Wait and See Season	79
8	Farm Bill Preliminaries Nonetheless	85
9	Negotiations in the Senate	95
10	House Approval of Freedom to Farm	105
11	At Last, a Farm Commodity Act	115
12	A Future for Farm Commodity Legislation	125
13	A Primer on U.S. Agricultural Policy Before 1996	133

Appendixes   141
  1. Senator Lugar's Questions on Prospective Farm Policy   141
  2. Early 1995 Anonymous Draft Approach to Withdrawal of Government Farm Programs   151
  3. The Summer of 1995 Anonymous Draft of "The Freedom to Farm Act of 1995"   155
  4. Chronology for the 1996 Farm Act   159
  5. Corn Prices and 1996 Farm Act Milestones   161

Notes   163

Bibliography   173

Index   177

Preface

This book is about the making of the 1996 Farm Act as perceived and understood by two observers. Any story reflects the experiences of its author (or authors, as in this case). One of the authors is a transplant from a sharecrop tenant family in the cornfields of Illinois to government in Washington. There he spent his professional career assisting in the analysis and implementation of U.S. Department of Agriculture (USDA) programs, observing policy making in Washington, and trying to understand changes in U.S. agriculture and the effects of farm policy on these changes. He also worked with the Senate Committee on Agriculture, Nutrition, and Forestry for a year, assisting in its preparation for the deliberations about the farm bill. The other author is urban born, a professor at a leading land grant institution, and a longtime student of trends and conditions characteristic of our society as a whole and of U.S. agriculture in particular. He is comfortable dealing with topics ranging from the political economy of agriculture and the history of U.S. farm policy to environmental issues and their relationships to agriculture.

What we came to understand about the making of the 1996 Farm Act is far from the complete story of the fashioning of the legislation. Our story surely omits important events and in other cases misinterprets them. Thus, we recognize in our work dimensions of Aesop's fable of the blind men and the elephant. Each individual observing or participating in the making of the farm legislation certainly "sees" some things differently than we do.

We deliberately focused primarily on the activities in Congress. It was more accessible, an important characteristic given the limits on our resources to track events, gather documents, interview participants, and draft text. In choosing this approach, we recognized that it would affect our view of the activities. The alternative of attempting to focus with equal diligence on all activities of all players in the making of the legislation would have dissipated our efforts, we think, too greatly. We also focused on commodity policies, giving less coverage to conservation and food policies.

Readers of a description of legislative activities impute an air of inevitability and even, in some cases, causality between events at one time and events at a later time; that is, what happened next must surely have been because of what occurred in the past. Sometimes this is true; sometimes it is not. As will be obvious to readers, there was no inevitability to the outcome for the 1996 Farm Act until near the end. We worked hard to avoid implying causality and inevitability when it did not exist, but we warn readers we may have slipped a few times.

Preparing this kind of book would be impossible without the help of a wide range of people—members of Congress and their staffs, lobbyists, administration officials, and competent newspeople—all of whom play critical roles in our democratic government. We appreciate their graciousness and kindness. In spite of our dependence on them, all errors in the book are our responsibility.

Lastly, this effort would not have happened without the support of the Economic Research Service of the USDA, an agency chartered to provide objective information and analysis for policy making and program implementation, as well as private decision making. Its cooperative agreement with Purdue University provided both the catalyst and the resources for the activities necessary in 1995 and 1996 for the gathering of the information critical to the writing of the manuscript.

Introduction

The story of how the 1996 Farm Act came to be is in many respects a story about Congressman Pat Roberts, R-Kans., who was the chairman of the House Committee on Agriculture in the 104th Congress and is now a senator. He, more than anyone else, shaped the legislation and thereby U.S. farm commodity policy for the years 1996–2002. His adroit maneuvering in the turbulent 1995 and early 1996 political happenings was key to shielding the federal transfers to commodity producers and farmland owners from the sharp budget-cutting knives of the House leadership.

The story of the 1996 Farm Act is also a story about Sen. Richard G. Lugar, R-Ind., the chairman of the Senate Committee on Agriculture, Nutrition, and Forestry. Early on, he proposed phasing out commodity programs. However, he was unable to attract sufficient support for his proposal for it to be voted on in the Senate. The two Democratic senators from North Dakota, other Democratic senators on the Senate Agriculture Committee, and most Republican senators from the principal farming states were simply not prepared to accept a phaseout of commodity programs. Confronted with this opposition Senator Lugar, in the end, accepted Congressman Roberts's approach—freedom to plant and guaranteed checks for seven years, 1996 through 2002.

Another legislator who played a key role was Sen. Patrick J. Leahy, D-Vt., the ranking Democrat on the Senate Agriculture Committee. He seized an opportunity to garner committee votes for his policy priorities—environment-related policies and approval of a northeast dairy compact that would increase the price paid for milk in the northeast—and gave in exchange his support of Congressman Roberts's approach to commodity legislation.

The 1996 Farm Act was passed by Congress in March 1996 and signed by the president on April 4, 1996 (Public Law [PL] 104–127). It established a new basis for deciding the size of the farm commodity check for any one person or entity. Further, it virtually guaranteed those federal payments regardless of forthcoming weather or market conditions. The critical new approach for determining who would receive farm commodity program benefits became, "Have you received federal farm commodity program checks in the past five years? If so, you are entitled to receive a check in each of the next seven years." This significant policy innovation is without precedent in U.S. farm policy history.

In a larger sense, however, the Federal Agriculture Improvement and Reform Act of 1996 (FAIR), the technical name of the 1996 farm legislation, continues the regressive transfers and reflects the continued political power of wealthier and higher-income U.S. farm interests.

The primary objective of this book is to record major developments associated with the formation and passage of the 1996 Farm Act. The congressional activities, including those related to the federal budget, are the special focus of this book. However, discussion of selective related activities of the Clinton administration and interest groups is also included. This record of these developments is important to any assessment of the prospects for continuing farm-related legislation into the future. A key question is whether the events of 1995 and 1996 related to farm commodity legislation are harbingers that the federal transfers to producers and farmland owners will end in 2002, or that the transfers will continue largely unabated over the years ahead.

We do not conclude from the 1995–96 experience that federal transfers to producers and farmland owners are likely to end in 2002, as some media writers seemed to want to believe as they reported on the passage of the 1996 Farm Act. Instead, we conclude the opposite: The federal transfers are very likely to continue. This conclusion is supported by Congressman Roberts's successful promotion of the policy eventually included in the 1996 Farm Act and by the support this policy received from other members of the House of Representatives and the Senate in spite of the expressed anguish over the federal budget deficit. Thus, we interpret the 1996 Farm Act history to suggest that members of Congress who embrace the interests of producers and farmland owners, although a minority, can be expected to skillfully craft farm program provisions that attract sufficient support in Congress to perpetuate transfers well into the twenty-first century.

Clearly, market realities constrain Congress in how to accomplish these transfers. Two important considerations are that U.S. farm production capacity greatly exceeds U.S. domestic demands and that prices of U.S. farm products are critical determinants of farm exports. The 1996 Farm Act provisions with respect to farm price supports most clearly demonstrate that a majority of farm legislators recognize these realities—realities that will prevail well into the coming century and surely will be the case when the 1996 Farm Act expires in 2002.

Acronyms

AAA	Agricultural Adjustment Act
AFBF	American Farm Bureau Federation
AFDC	Aid to Families with Dependent Children
ARP	Acreage Reduction Program
ASCS	Agricultural Stabilization and Conservation Service
CBO	Congressional Budget Office
CCC	Commodity Credit Corporation
CFSA	Consolidated Farm Service Agency
CRP	Conservation Reserve Program
CRS	Congressional Research Service
EEP	Export Enhancement Program
EQIP	Environmental Quality Incentive Program
FSA	Farm Service Agency
FTF	Freedom to Farm
FY	Fiscal Year
GATT	General Agreement on Trade and Tariffs
MPP	Market Promotion Program
NRCS	Natural Resources and Conservation Service
OMB	Office of Management and Budget
PL	Public Law
REA	Rural Electrification Administration
TAB	Total acreage base
USDA	U.S. Department of Agriculture
WIC	Special Supplemental Nutrition Program for Women, Infants, and Children
WRP	Wetlands Reserve Program

SENATORS AND REPRESENTATIVES

SENATORS

Akaka, Daniel K., D-Hawaii
Baucus, Max, D-Mont.
Breaux, John B., D-La.
Bumpers, Dale, D-Ark.
Cochran, Thad, R-Miss., Chairman, Committee on Appropriations, Subcommittee on Agriculture, Rural Development and Related Agencies and Chairman, Committee for Agriculture, Nutrition, and Forestry, Subcommittee on Production and Price Competitiveness
Conrad, Kent, D-N.Dak., Ranking Democrat, Committee for Agriculture, Nutrition, and Forestry, Subcommittee on Marketing, Inspection, and Product Promotion
Coverdell, Paul, R-Ga.
Craig, Larry E., R-Idaho, Chairman, Committee for Agriculture, Nutrition, and Forestry, Subcommittee on Forstry, Conservation, and Rural Revitalization
Daschle, Thomas A., D-S.Dak., Senate Democratic Leader
Dole, Robert, R-Kans., Majority Leader
Domenici, Pete V., R-N.Mex., Chairman of the Senate Committee on the Budget
Dorgan, Byron L., D-N.Dak.
Exon, J. James, D-Nebr.
Feingold, Russell D., D-Wis.
Ford, Wendell, D-Ky.
Glenn, John, D-Ohio
Grassley, Charles E., R-Iowa
Harkin, Tom, D-Iowa, Ranking Democrat, Committee for Agriculture, Nutrition, and Forestry, Subcommittee on Research, Nutritiion, and General Legislation
Hatfield, Mark O., R-Ore., Chairman, Committee on Appropriations
Helms, Jesse, R-N.C., Chairman, Committee for Agriculture, Nutrition,

and Forestry, Subcommittee on Marketing, Inspection, and Product Promotion
Inouye, Daniel K., D-Hawaii
Johnston, J. Bennett, D-La.
Kerrey, J. Robert, D-Nebr.
Kohl, Herb, D-Wis.
Leahy, Patrick J., D-Vt., Ranking Democrat, Committee for Agriculture, Nutrition, and Forestry
Lott, Trent, R-Miss., Assistant Republican Leader
Lugar, Richard G., R-Ind., Chairman, Committee for Agriculture, Nutrition, and Forestry
Mack, Connie, R-Fla.
McCain, John, R-Ariz.
McConnell, Mitch, R-Ky., Chairman, Committee for Agriculture, Nutrition, and Forestry, Subcommittee on Research, Nutrition, and General Legislation
Nunn, Sam, D-Ga.
Pryor, David, D-Ark., Ranking Democrat, Committee for Agriculture, Nutrition, and Forestry, Subcommittee on Production and Price Competitiveness
Santorum, Rick, R-Pa.
Thurmond, Strom, R-S.C.
Warner, John W., R-Va.
Wellstone, Paul, D-Minn.

Representatives

Armey, Richard K., R-Tex., Majority Leader of the House
Baker, Richard, R-La.
Barrett, Thomas M., R-Nebr., Chairman, Committee on Agriculture, Subcommittee on General Farm Commodities
Bishop, Sanford D., D-Ga.
Boehlert, Sherwood L., R-N.Y.

Chabot, Steve, R-Ohio
Chambliss, Saxby, R-Ga.
Condit, Gary A., D-Calif., Ranking Democrat, Committee on Agriculture, Subcommittee on Department Operations, Nutrition, and Foreign Agriculture
DeLay, Tom, R-Tex., Republican Whip
de la Garza, E, D-Tex. Ranking Democrat, Committee on Agriculture
Durbin, Richard J., D-Ill., Ranking Democrat, Committee on Appropriations, Subcommittee on Agriculture, Rural Development, Food and Drug Administration, and Related Agencies
Emerson, Bill, R-Mo., Chairman, Committee on Agriculture, Subcommittee on Department Operations, Nutrition, and Foreign Agriculture
Combest, Larry, R-Tex.
Frank, Barney, D-Mass.
Gingrich, Newt, R-Ga., Speaker
Gunderson, Steve, R-Wis. Chairman, Committee on Agriculture, Subcommittee on Livestock, Dairy, and Poultry
Kasich, John, R-Ohio, Chairman, Committee on the Budget
Kennedy, Joseph P., D-Mass.
Livingston, Bob, R-La., Chairman, Committee on Appropriations
Lowey, Nita M., D-N.Y.
Miller, Dan, R-Fla.
Peterson, Collin C., D-Minn.
Roberts, Pat, R-Kans. Chairman, Committee on Agriculture
Rose, Charlie, D-N.C.
Schumer, Charles E., D-N.Y.
Shays, Christopher, R-Conn.
Skeen, Joe, R-N.Mex., Chairman, Committee on Appropriations, Subcommittee on Agriculture, Rural Development, Food and Drug Administration, and Related Agencies
Solomon, Gerald B., R-N.Y., Chairman, Rules Committee
Stenholm, Charles W., D-Tex., Ranking Democrat, Committee on Agriculture, Subcommittee on General Farm Commodities
Zimmer, Dick, R-N.J.

The Making of the 1996 Farm Act

Five Early Developments: April 1994 to April 1995   1

Five developments between April 1994 and April 1995 were especially relevant to the efforts to craft what was eventually the 1996 Farm Act. They were

—The preparation and release of a consultant's study entitled *Large-Scale Land Idling Has Retarded Growth of U.S. Agriculture.*
—The Republicans gaining control of both houses of Congress.
—Sen. Richard G. Lugar's, R-Ind., early agenda for a debate over farm legislation.
—An anonymous January 1995 three-page paper describing one approach to extricate the federal government from farm commodity programs.
—The subordination of the Senate Committee on Agriculture, Nutrition, and Forestry and the House Committee on Agriculture to the authority of the Senate and House Committees on the Budget during the first year of the 104th Congress.

The *Large-Scale Land Idling Has Retarded Growth of U.S. Agriculture* study marked the initiation of a successful, organized lobbying effort to get the government out of yearly determinations of land area to be withheld from producing the major crops including corn, cotton, rice and wheat.

With the Republican control of both houses of Congress, the chairmanship of the House Agriculture Committee was assumed by Congressman Pat Roberts, R-Kans., and the chairmanship of the Senate Agriculture Committee by Senator Lugar. Thus, the Republican party leadership had central control of the agenda of both Agriculture Committees.

Senator Lugar's agenda included a vigorous dialog about farm policy. Shortly after the 1994 national election he called for a discussion about the purposes, effectiveness, and utility of farm programs, including those directly related to commodities. The seriousness with which different groups took this challenge varied greatly. In the end, the aspirations for an objective dialog over farm policies were submerged by the more typical struggles over votes on legislative provisions that would accommodate budget dictates, but nonetheless substantially continue the flow of federal money to producers and farmland owners.

The January 1995 anonymous three-page paper described a policy approach that would reduce farm commodity program payments to producers and landowners year by year. The approach described in this paper is strikingly similar to the content of another anonymous three-pager that appeared in the summer of 1995. Its features became the core provisions of the freedom to farm proposal and Title I of the 1996 Farm Act.

The Republican leadership's concern about the federal budget deficit, together with budget procedural rules of the Congress, was quickly reflected in the dominance of the Budget Committees over the agendas and activities of other committees, including the Agriculture Committees. Specifically, the Agriculture Committees were challenged to develop farm legislation that comported to budget disciplines as defined by the Budget Committees and particularly by the chairmen of these committees.

THE RELEASE OF LARGE-SCALE LAND IDLING HAS RETARDED GROWTH OF U.S. AGRICULTURE STUDY

The idea that farm programs had gone too far in withholding cropland from production was given a substantial boost with the preparation and astute promotion of a study sponsored by the National Grain and Feed Association through their foundation. The study, released in May 1994, concluded that "American farmers and the U.S. economy stand to reap substantial benefits from expanding crop area and production."[1] Over 185 companies, most of whose profits are geared substantially to volume of commodities handled or processed, were involved in supporting the study prepared by Abel, Daft, & Earley, a consulting firm in the Washington, D.C., area.

Principal Finding

The key conclusion of the study was that 38 million of the then 65 million acres of cropland held out of production at that time under the

Acreage Reduction Program (ARP), the Conservation Reserve Program (CRP), and other, but smaller, programs could, under expected demands and yields, be brought back into production between 1994 and 2002 and commodity prices *would not be less than they were at the time of the study*. Politically, that is a powerful conclusion for there is a strong preference among politicians not to be accused of taking action which leads to lower producer prices. Central to this proposition was the conclusion that demands for U.S. farm commodities would increase enough so that farm commodity prices in the prospective future would not drop below then current levels, even if U.S. farm production increased as hypothesized. The implication for farm income was obvious—more production at the same or higher prices meant more income.

The study also concluded that the cultivation of an additional 38 million acres would generate additional employment. Thus, the study addressed an additional policy issue, job creation, which was receiving political attention in 1994.

The study was good news for its sponsors, who tend to operate on relatively fixed margins and whose profits are therefore tied substantially to volume. The text of the report did not discuss this volume-profit situation for agribusinesses. Nonetheless, the study skillfully exploited the general concept that profits rise if fixed costs can be spread over larger volumes, as it reasoned why producers would benefit from using more of their cropland in production rather than withholding it.

Things Not Said

There were two pieces of information that, if they had been included, would have made the study more informative and thereby enhanced the quality of the coming debate over farm legislation. One piece that would have added important insights would have been estimates of federal commodity program outlays with an additional 38 million acres in production (the proposed expanded U.S. cropland acreage scenario). The other piece would have been a comparison between estimated farm prices and income with the additional 38 million acres in production scenario emphasized in the study, and estimated farm prices and income if U.S. crop acreage were not increased.

Although the study did not include estimates of federal outlays associated with the expanded cropland acreage scenario, individual farm worksheets used in meetings in which the study was discussed with producers showed higher government farm program deficiency payments with the expanded acreage options ("zero ARP") than with a no policy change option ("10% ARP").[2] These differences indicate that the authors expected that farm commodity market prices would be lower under the expanded U.S. cropland acreage option *than they would be under the no policy change option,* and logically so. One possible explanation for the study not including total outlay/budget estimates is that the importance the 104th Congress would attach to budget matters, specifically outlays, was not anticipated. Another possible explanation is that the sponsors or the authors of the study concluded that attention to budget matters involving increased commodity program spending would detract from the study's intended message.

Inclusion of the second piece of information—a comparison of farm prices and incomes for two future scenarios—would have contributed even more to the farm legislation debate than the inclusion of outlay estimates. The study focused on the United States as a whole and compared what the experience *in the past* had been under a regime that in recent years withheld 65 million acres from production with what the experience *in the future* (2002) would be under a regime that allowed 38 of the 65 million acres to come back into production.

This type of comparison between situations in two different time periods can be misleading. People may be induced to think that the change in policy that is highlighted—in this case allowing 38 million acres of cropland withheld in the 1994–95 years from production to come back into production in subsequent years—accounts for or causes the contrasting outcomes in the two different time periods when it may not. This confusion could have been overcome by including another set of estimates—namely what the experience in the future (2002) would be with a scenario that did not allow any of the 65 million acres to come back into production.

Regardless of how the study might have made a greater contribution to enlightened debate, it is clear that the leaders of the National Grain and Feed Association astutely anticipated an important opportunity to press for expanded U.S. production without appearing to be opposed to the economic welfare of U.S. producers. As the farm bill debate unfolded, the drive for less regulation on producers was translated to include an appeal for more acreage in production. Also, as it turned out, commodity prices continued to strengthen through 1995 and into 1996, thus reinforcing the political momentum for the federal government to relax the restraints on the amount of cropland used to produce crops. This market development fit perfectly with the mood of the new free market-oriented Republican Congress.

The Republicans Gain Control of Congress

With the November 1994 elections, the Republicans gained control of Congress. The drive of the new congressional leadership to develop a plan to balance the federal budget meant that Agriculture Committee matters were overwhelmed by budget procedures and directives. Committee leaders had limited time to develop approaches to farm commodity policy that would attract support of Democratic as well as Republican members of their committees. In the past the development of farm commodity legislation often involved the finding of common ground among both Democrats and Republicans that would modify, but nonetheless continue, the basic approaches to effecting income transfers to producers and farmland owners. However, that approach could no longer be assumed, especially with the Republican congressional leadership seemingly tied very closely together to assure the development of a plan to balance the federal budget.

Different People in Leadership Positions

With the Republican Party gaining control of both the Senate and the House, the chairmen of the Agriculture Committees changed. In the Senate, Senator Lugar replaced Patrick J. Leahy, D-Vt., as chairman of the Agriculture Committee, and Thad Cochran, R-Miss., replaced Dale Bumpers, D-Ark., as chairman of the Committee on Appropriations, Subcommittee on Agriculture, Rural Development, and Related Agencies. In the House of Representatives, Congressman Roberts replaced E. de la Garza, D-Tex., as chairman of the Agriculture Committee, and Joe Skeen, R-N.Mex., replaced Richard Durbin, D-Ill., as chairman of the Committee on Appropriations, Subcommittee on Agriculture, Rural Development, Food and Drug Administration, and Related Agencies.

Changes in committee staffs accompanied the changes in the leadership of the committees. Charles F. Conner, who had worked as minority staff director, became majority staff director of the Senate Agriculture Committee. In the House, Gary Mitchell became the majority staff director for the Agriculture Committee. Like Connor in the Senate, Mitchell had worked as minority staff director in previous sessions of Congress when the Democrats were the majority.

Uncertainty about Armey's Role

The election of Richard K. Armey, R-Tex., as the majority leader of the House was potentially more important for farm legislation than the change in chairmen of the Agriculture Committees. Congressman Armey had been critical of farm commodity programs over many years, and the majority leader position provided important opportunities to influence farm legislation. A big question was the extent to which Congressman Armey would use the position to curtail and limit the farm commodity programs he disliked so intensely.

Importance of the Contract with America to Farm Legislation

The "contract with America," a major fixture during the election campaign and the early months of 1995, did not include the word "agriculture" or the word "farming," nor did the contract identify farm commodity programs. Nonetheless, the contract had significant implications for the ensuing jockeying over farm commodity programs. The most notable implication was how the contract committed the Republicans to developing a plan for balancing the federal budget and its underlying theme of reducing the role of the federal government in the economy. This commitment led to a series of steps that forced the Agriculture Committees to seriously examine ways that would generate the budget savings required by the concurrent budget resolution managed by the Budget Committees.

The Republican control of the House was linked directly to the election of seventy-three freshman Republicans. Their attitudes toward restraining government outlays and reducing the role of government were well known. Each of them had signed the "contract with America." What was not known, however, was whether they would make a distinction between welfare for poor people and transfers to farm producers and farmland owners. Several signers (nearly one-third) represented congressional districts with substantial amounts of agricultural production.

Comments by Earl Butz during U.S. Department of Agriculture (USDA) agency budget discussions while he was secretary of agriculture illustrate the importance of the distinction among recipients of federal outlays. During one series of budget review sessions with his agency administrators, Secretary Butz used a four-by-six-foot chart displaying two lines. One depicted the increasing outlays over time for food assistance programs, like food stamps, in the 1960s and into the 1970s. The other line depicted the then declining outlays associated with farm-related programs that generated U.S. government checks to farm producers and farmland owners. He first pointed to the food program outlay line and said in a disparaging tone, "That line represents welfare and it keeps increasing." With heads around the table nodding and thereby silently endorsing the unspoken notion that it was undeserved, Secretary Butz pointed to the declining line for farm-related programs and stated, "Look at the decline in this line! Now, it is welfare, too. But it is for our kind of people!"

Whether the seventy-three freshman Republicans would distinguish between the two kinds of welfare was unknown. Also, yet to be revealed was how Congressman Roberts would balance the budget savings demands placed on him by the House leadership with the interests of traditional supporters of farm commodity programs. It became increasingly evident that he would strongly defend farm program outlays and accept Budget Committee directives only after vigorously contesting them.

One of the major implications of the Republican control of the House and the primacy the party gave to balancing the budget was the dominant role that the Budget Committee would exert on all legislation. The Agriculture Committee quickly became unable to continue the long bipartisan tradition of formulating farm commodity policy that attracted both Democratic and Republican support. Instead, the forced responses to the Budget Committee left little time to find bipartisan common ground on any policy, much less policies that generated Congressional Budget Office (CBO) estimates of budget savings. The tensions related to budget matters even led to difficulties among Republican members of the House Agriculture Committee. It was not uncommon to hear that Democratic members were not invited to House-Senate conference committee meetings. But on occasion selected Republicans were not invited either.

Senator Lugar's Early Agenda for the Farm Bill Debate

In a manner very different from the defensive approach of the House Agriculture Committee's leadership, one of Senator Lugar's first activities as chairman of the Senate Agriculture Committee was to let it be known that he wanted to consider major changes in farm commodity programs. He had his staff develop a series of questions about the programs under the jurisdiction of his committee and the "underlying assumptions that have shaped these programs over the course of many decades."[3] These questions were released to the press and were circulated widely among those interested in commodity programs (see Appendix 1).[4] Senator Lugar solicited responses from farm groups to the questions. In his letter to them he argued that it was appropriate to review these programs not only because of budget pressures but also because "U.S. agriculture is changing, while the programs designed to serve it have, in many cases, not changed." He observed that "every agricultural program will be called on to justify its existence and continuation."

Senator Lugar's approach was not typical for a Midwest farm state senator. However, it was consistent with his reputation of being willing to open a good faith broad public debate over issues about which he felt strongly. His proposal to consider major changes in the programs was criticized in the Midwest, although the Indiana Farm Bureau supported his approach.

A Letter to the Chairman of the Senate Committee on the Budget

On December 1, 1994, Senator Lugar sent a letter to Sen. Pete V. Domenici, R-N.Mex., chairman of the Senate Budget Committee, in which he stated that he planned for the Agriculture Committee to "conduct a thorough 'bottom-up' review of all farm programs." He went on to state, "At this point, I would not rule out any options, including the abolition of the programs, their conversion into block grants for administration by the States, the creation of a user-funded revenue assurance program, or a redirection of funds into priorities like research, rural empowerment and market development." At the same time, he made clear his discouragement with the inability to accurately anticipate the budgetary outlays associated with farm programs and he called for limiting their entitlement status by stating, "We must reform agricultural programs in such a way that spending has limits in which we and all citizens can have confidence."[5]

Note that Senator Lugar spoke about limits; he did not suggest that the amount of transfers be fixed. Later, however, he played a critical role in Congress's embracement of a farm commodity program that would transfer fixed annual amounts to commodity producers and farmland owners—the Freedom to Farm program as crafted by Congressman Roberts in the House and modified in the House-Senate conference on the budget reconciliation legislation. That particular 1995 budget legislation was later vetoed by the President. However, its major agricultural provisions were later approved by both the Senate and the House as part of the 1996 Farm Act.

A Specific Proposal

Senator Lugar also announced a specific proposal on how the wheat, feed grains, oilseeds, rice, and upland cotton programs should be adjusted. He proposed that the target prices of these crops be reduced by 3 percent per year. This approach was one of two ways frequently discussed to adjust commodity programs to realize lower federal outlays. The other was to reduce the acreage to which federal deficiency payments applied.

Comparisons

To understand Senator Lugar's specific proposal and how it compares to reducing the number of acres for which deficiency payments were made, it is important to understand the interaction among target prices, set-aside or flex acres, and government outlays in the pre-1996 farm programs. The deficiency payments for any one participant in the program for any one particular crop under the 1990 farm legislation were equal to the product of three items:

—The payment amount, which was the difference between the national target price and market prices as determined by the USDA (or the loan rate if it was higher than the market price).
—The payment yield, which was the "established crop yield" for particular parcels of land included in the USDA records.
—The payment acres, which was the base acreage for the individual farm carried in the USDA records for that farm reduced by a nationally set determined set-aside or flex acreage percentage.

The payment amount per bushel times historical yield times acres qualifying for payments equaled the amount paid to those participating in the program.

Thus, a reduction of national target prices, payment yields, or payment acres would, under the pre-1996 program, reduce outlays for deficiency payments. However, the outlay effects among these three posited alternatives would vary as would the distribution effects among producers. For example, a 1 percent cut in the established payment *yields* for all producers for all program crops, or a 1 percent cut in the national payment *acres,* would lead to the same percentage cut (1 percent) in deficiency payments that any one individual producer of any of the supported crops would receive from the federal government. With these approaches deficiency payments for all producers for all program crops would be cut proportionally relative to what the deficiency payments would be without the cuts.

In contrast, consider a situation whereby national *target prices* were adjusted downward to achieve budget savings and equal percentage cuts were applied to the target prices of all program commodities. Then, the relative effects on the program benefits among producers of different commodities could differ substantially depending on the ratio of market prices to target prices across commodities at that time. Nominal benefits could, with this approach, become more equal or less equal. Suppose, for example, that wheat had a target price of $4.00 and a market price of $3.40, and rice had a target price of $10.71 but a market price of $6.50. Then for every 1 percent cut in the target price of wheat, the deficiency payment would be nearly 7 percent lower. But for rice a similar cut in the target price of 1 percent would translate into approximately 2.5 percent lower deficiency payments. Thus, changes in target prices by an equal percentage would result in different relative changes among commodities in payment rates to producers and landowners. The numbers in Table 1.1 related to the 1993 crop year place these relationships into perspective.

The acceptance of Senator Lugar's suggestion to reduce target prices by 3 percent per year would have led to the demise of deficiency payments, but the lengths of time for the payments to disappear would differ among commodities. If the 1993 commodity market prices remained the same in the future, corn producers would cease receiving payments in about four years, and payments to rice producers would not cease until nine years.

Obviously, there are several ways in which the government might discontinue support to commodity producers. Senator Lugar's specific proposal to reduce target prices by 3 percent is just one that would have been consistent with his statement on National Public Radio on January 13, 1995, that, "I believe that we ought to move the federal government out [of farm commodity programs] altogether."[6] However, Senator Lugar's proposal raised real concerns among commodity groups about disparity of treatment, which became a rallying cry against any cuts at all.

TABLE 1.1. Effects of cutting commodity target prices based on data for the 1993 crop

Commodity	Target price	Market price	Deficiency payment	Effect of a 3% reduction in target prices on the deficiency payments[a]	
			dollars		percent
Corn (per bu)	2.75	2.47	0.28	−.825	−29
Rice (per cwt)	10.71	6.73	3.98	−.321	− 8
Upland cotton (per lb)	0.7290	0.543	0.1860	−.2187	−12
Wheat (per bushel)	4.00	2.97	1.03	−.120	−12

[a]Assume that changes in deficiency payments are equal to dollar equivalent of 3% decreases in target prices.

THE ANONYMOUS JANUARY 1995 PAPER ON A TEN-YEAR TRANSITION

In the middle of January 1995 a three-page paper and a related one-page paper entitled "Ten Year Transition Period from the Present Commodity Price Support, Marketing Loan and Deficiency Payments" circulated among a small group of Senate Agriculture Committee staffers and USDA and land grant university agricultural economists (see Appendix 2).[7] Both were anonymous and part of the materials brought together in preparation for a weekend meeting of the participants to discuss possible approaches to eliminating or radically changing farm commodity programs.

This January 1995 three-pager described a ten-year transition period during which those receiving deficiency payments over the past three years would be eligible to receive checks. Their deficiency payment checks in the first transition year, according to this paper, would be 90 percent of the average they received in the past three years. Similarly, those who received nonrecourse loans would be eligible to obtain nonrecourse loans. In subsequent years, according to the three-pager, the deficiency payment checks would decrease 10 percent per year, and price support levels would be reduced 10 percentage points in each year.

The entitlement to the new deficiency payment checks would be partially attached to the land and be transferable with land. Another portion would be attached to the "individual operator [sic] of the land in 1992–94" and would presumably not be transferable. The proportion attached to the land would initially be high and that attached to individuals would be low. However, the proportion attached to the land would incrementally decrease and that to operators would incrementally increase. This feature was included to confront the longstanding problems associated with farm commodity program benefits being capitalized in land prices. The resulting higher land prices have benefited the landowners, but the benefits of the programs to operators of farmland who did not own the land they operated were dissipated through higher rents for the higher-priced land. The intent here was to tilt the payments more in favor of operators as contrasted with landowners who have been the primary beneficiaries of past commodity programs.

The January 1995 three-pager, as well as the related one-pager, were held closely. However, it is intriguing to compare the approach described in these papers with the content of another three-pager, also anonymous, that emerged some four months later, circulated widely among congressional staff and interest groups, and then formed the basis for Congressman Roberts's Freedom to Farm proposal, and in the end the 1996 Farm Act (see Appendix 3).[8]

It is also useful to note that John Baize, an individual who had worked for commodity organizations for several years, elaborated concepts in 1985 that were considered at the time to be heretical, but included features that would have led to a gradual phaseout of the programs. The programs as they were operating in the mid-1980s would be replaced, following the Baize concepts, with two features. Operators/owners of each acre of cropland would receive fifty dollars for each acre in the first year of his program. This payment would decline to zero in twenty years. Producers and landowners receiving commodity loans would be required to repay the loans. They could no longer give the government a related quantity of the commodity as settlement for commodity loans. In addition, the loan levels expressed in dollars and cents per unit of production would be low relative to usual market prices. Its simplicity is disarming, but its transparency was probably one of the important reasons that the approach was not embraced. For example, one senator reportedly said, "You have to make it complicated. If they understand it, they may not buy it." That quotation helps one understand why the U.S. farm commodity programs have become so complicated and perhaps why the 104th Congress's House leadership became so exasperated with the difficulty they experienced in dealing with farm legislation.

SENATE AND HOUSE BUDGET COMMITTEES AFFIRM THEIR AUTHORITY

By the end of April 1995 it was evident that the Republican leadership was insistent that the Congress develop and vote on a fiscal plan that would reflect a balancing of the federal government budget within a few years. Further, to accomplish this feat they were prepared to follow the procedures incorporated in laws passed by earlier congresses. Consequently, the authorities of the Budget Committees were reaffirmed. Authorization committees, like the House Agriculture Committee, and the appropriations committees and their subcommittees, like the House Agriculture Appropriations Subcommittee, saw their agendas dictated by the Budget Committees and their actions constrained by instructions from these committees in terms of how much calculated budget savings their changes in legislation were to generate.

Other Critical Developments from April 1994 to April 1995

There were other developments in 1994 and early 1995 that were relevant to prospective commodity legislation. These ranged from actions by the administration which had spillover effects on later farm bill activities to debates among people who viewed the need for federal involvement in farm commodity programs in vastly different perspectives to studies, reports, and surveys which added to the available information important to preparations for a vigorous debate about the direction that farm commodity legislation should take.

Promises to Assure Passage of Trade Legislation

When pressing in September 1994 for passage of the General Agreement on Trade and Tariffs (GATT) implementing legislation, Secretary of Agriculture Mike Espy and Alice Rivlin, the acting director of the Office of Management and Budget (OMB), sent a letter dated September 30, 1994, to the chairmen and the ranking minority members of the Agriculture Committees. This letter contained assurances of support for export subsidy programs; the expansion of the purposes of these programs; a "full continuation" of the CRP, which withheld land from commodity production through annual payments to its owners; maintenance of discretionary spending on USDA agricultural programs; and increases of $600 million in GATT-allowed "greenbox" and other activities to promote exports of U.S. farm products.[9] A companion letter signed by President Clinton reaffirmed the substance of the Espy/Rivlin letter.[10]

Obviously, this correspondence was meant to attract votes for the pending trade legislation. In one sense the letters contributed clarification to how the administration was interpreting the provisions of the trade agreements that had been negotiated and were being considered by Congress. But in a larger sense the need for votes for the trade legislation provided an opportunity to extract from the administration the letters and promises to continue particular transfers to the farm sector. The letters strengthened the position of those in Congress and in the administration who favored a continuation of the income transfers to agriculture.

Administration Farm Budget Hawks and Doves

In September–October 1994 Alice Rivlin and her staff prepared an October 3, 1994, "Big Choices" memorandum. The memorandum was marked "for handout and retrieval in meeting" and was presumably prepared for use in a meeting of high-level members of the executive branch. It argued that "decisions must be made soon about the policies to be articulated in the FY1966 budget, the State of the Union, and our response to the Kerrey-Danforth Commission report." It identified five goals that would involve new federal outlays and/or lower federal revenues. In addition, it questioned where the resources for these initiatives would come from. The list of "Illustrative Entitlement Options" for finding these resources included, "Reduce agricultural target prices by 3% per year ($12B) and increase triple base acreage to 25% ($4B) for a combined saving of $16 billion."[11]

Unofficial reports indicated that President Clinton was very critical of these types of proposed cuts in farm programs, which also came from the Council of Economic Advisers. He reportedly exclaimed during one of the exchanges that no one except himself understood agriculture. One individual indicated that she came the closest to ever being fired when she advocated to the president that farm program outlays should be cut by substantial amounts.

Clashes Between Senator Lugar and the Democrats

Another development was the public clash between Senator Lugar and the Democrats on the Senate Agriculture Committee as illustrated by their presentations to the Senate Budget Committee on February 16, 1995. In his testimony before the committee Senator Lugar noted Budget Committee Chairman Domenici's challenge to committee chairmen "to spell out how federal spending in their areas of responsibility ought to change." Senator Lugar proposed saving "almost $15 billion over five years in two simple steps. First, we should save $11.45 billion by reducing target prices on the major program crops by 3% a year, each year, for five years, with corresponding reductions for those commodities that are subsidized in ways other than direct payments. Second, we should save $3.4 billion by eliminating the Export Enhancement Program."[12]

Senator Lugar went on to explain that, in the past, hypothetical farm program savings did not materialize. He pointed to the belief expressed in 1990 that the 1990 Farm Bill would reduce outlays for the programs by $10 billion over the five-year life of the bill. Instead, the outlays were $15 billion more than anticipated over the bill's life.

Senator Lugar's proposal would, if implemented, reduce outlays from what they would be otherwise. However, his specific proposal to reduce target prices did not deal directly with the concern about farm program outlay uncertainty that he had been emphasizing. Admittedly, the uncertainty associated with total farm program outlays would become mute as outlays approached zero.

In a rebuttal to Senator Lugar, Sen. Kent Conrad, D-N.Dak., a member of the Senate Agriculture Committee, argued that federal farm programs are justified because of their effect on international competitiveness, American consumers, the number of American jobs, and the production of food and fiber. He went on to argue that agriculture is not like "every other type of business" and stated that "agriculture is unique in supplying a basic necessity of human life under unstable weather conditions." He endorsed change by suggesting (1) doing more to negotiate "international trade agreements ... to level the playing field," (2) "work to harmonize our agriculture policies with other nations," (3) "promote value-added processing through the formation of cooperatives," and (4) "examine targeting of farm program benefits to efficient family-size producers."

Senator Conrad's only concession to budget stringency in his statement before the Budget Committee was his proposal to consider targeting program benefits to "efficient family-size producers." He recognized the "complexities of this issue" (targeting payments) and stated, "... but I believe we must—as part of the budget process—make difficult choices about federal farm program payments." The implication was that targeting payments to "family-size" producers would reduce payments to larger farms and thereby generate significant savings in federal outlays.[13]

The contrast between the two senators' views of (1) the characteristics of the farm sector, (2) the justification for continued federal transfers, and (3) the extent to which budget considerations should drive the decisions was stark and foreshadowed future legislative battles in the House as well as the Senate. These conflicts were to be among commodity interests, between Democrats and Republicans, among Democrats, and among Republicans—even among Republican members of the Senate Agriculture Committee.

Studies and Reports

Books and reports of various study groups focused on the prospective farm bill were released during this initial phase. These publications included a book prepared under the auspices of the North Central Regional Research Project.[14] Most, but not all, of the forty-five authors were affiliated with U.S. land grant colleges of agriculture. The authors appropriately described current policies and identified issues that were then expected to be central to the development of a 1995 Farm Bill.

These chapters reflected the expectation that the 1995 bill would constitute a further evolution of traditional farm policy. Three of the twenty-one chapters dealt with commodity policy issues, options in dealing with the issues, and the implications of these options. Although these chapters reflected, for the most part, an expectation that 1995 commodity policy changes would be evolutionary, the challenges to continuation of the sugar, honey, wool, and mohair policies were recognized. The approach Senator Lugar proposed for scaling back commodity target prices was not mentioned. Neither was the approach that became central to Congressman Roberts's Freedom to Farm proposal. The primary focus was on marginal changes in existing programs.

Also, the first of a series of linked USDA publications was released in April 1995. The 1995 reports were presented as updates of a series of reports released in anticipation of the 1990 Farm Bill. They included several data series and "summarized the experiences with various farm programs and the key characteristics of the commodities and the industries that produce them."[15] Most of the reports included a section on policy issues. Reference to options like Senator Lugar's and Congressman Roberts's proposals were limited. However, the wheat report discussed the possibility of "total flexibility" for producers to use their land for any commodity without limitations based on past use of the land; the feed grain report discussed modifying the approach to acreage idling within a context of continued deficiency payments tied to market prices; and the cotton report mentioned the possibility of changing target prices, loan rates, and the number of acres held out of production. However, as with the "land grant book," nothing involving as much change as Senator Lugar's or Congressman Roberts's proposals for farm commodity legislation was given significant attention.

In March 1995 the National Center for Food and Agricultural Policy released a series of reports of six working groups the center had organized with the cooperation of the Hubert H. Humphrey Institute of Public Affairs. Each of the working groups had been organized to focus on a particular subject expected to be a critical part of the prospective 1995 Farm Bill. One was focused on the long-standing policy topic "Price and Income Stability."[16]

As was the case for the other working groups in the center's effort, the Price and Income Stability Working Group was conceived as a way whereby people of different policy persuasions could come to better understand the issues, alternative approaches to the issues, and the implications of these alternative approaches. The purpose was not to develop advocacy for any particular program. With this planned approach, the "real" benefit of the activity would be seemingly unnoticed but reflected in activities of working group members as they individually engaged in the policy making process during 1995. Thus, an emphasis was placed on inviting people to be working group members who embraced different policy positions and were expected to be active in the 1995 farm policy debate.

In retrospect, there are probably several reasons why the planned nonadvocacy approach was not sustainable. The activities quickly shifted toward working group members individually advocating particular approaches. Members were assertive in expressing support or rejection of alternative approaches.

The report of the Price and Income Stability Working Group indicates that some working group members favored an end to commodity programs. However, possible approaches to ending commodity programs were not identified; therefore, the implications of following Senator Lugar's approach, or any other approach that ended the programs, were not discussed.

Although it does not appear to have confined the group, the title given to the working group, "Price and Income Stability," is not policy neutral. It set stability up as "good" and implicitly placed approaches like Senator Lugar's and Congressman Roberts's beyond the purview of the group. The general belief has been that government farm commodity program features like price supports, deficiency payments, and acreage retirement programs enhance stability and that ending them would induce greater instability of prices and incomes.

Overall, given the past evolutionary nature of farm policy in most periods, it is not surprising that the studies and pre-farm bill task forces stuck close to examining approaches that represented modest adjustments in policy. The thinking of Senator Lugar, and eventually Congressman Roberts, was much more revolutionary than evolutionary. An important question is whether serious analysis of revolutionary changes might have informed the debate, and if so would the resulting policy have been any different? Variations of this question include: Do studies and reports that focus on evolutionary approaches to policy help or hinder the efforts of people when they consider substantial change as did Senator Lugar and Congressman Roberts in 1995? Would society's needs be better served if more revolutionary approaches were analyzed more frequently in advance of policy decisions?

A July 1994 Senate Letter

Before the farm bill debate got underway, in July 1994 Senators Leahy and Lugar, then chairman and ranking minority member of the

Senate Agriculture Committee, wrote to one thousand people, who because of their responsibilities, had a substantial stake in "food and farm" legislation. They asked the addressees to identify a limited number of "specific changes" and/or "new approaches" to the related legislation and programs that the addressees thought "should be examined as the 1995 Farm Bill is considered by the Congress." Most of the 135 responses received suggested limited adjustments in this or that program. However, a very small number of respondents called for widescale changes. Two argued that the commodity programs should be eliminated or revised dramatically—"over the next five years," according to one, and "as rapidly as possible," according to the other. Another respondent proposed lowering grain target prices by twenty-five cents per bushel—an amount much less than what Senator Lugar later proposed. Dramatic changes in commodity programs would be involved if the suggestion of another response were followed. It called for linking program benefits to the income of recipients rather than according to a measure of production as past programs have done and as the programs pursuant to the 1990 Farm Act had done. Some respondents proposed fewer constraints on the use of farmland for crop production. Others proposed continuation and, in some cases, expansion of the CRP.[17]

In summary, in late 1994 and early 1995 only a limited number of people were thinking in terms of dramatic changes in farm commodity programs. The type of transformations in commodity programs that Senator Lugar soon advocated and Congressman Roberts later proposed were, with important exceptions, seldom identified and even less frequently seriously considered in the pre-farm bill preparations. This meant that analyses focused on the effects of major structural changes in commodity programs—like lowering target prices by 3 percent for each of ten years, targeting benefits according to need, or substituting fixed payments for payments geared to market prices—were simply not available. Thus, congressional staff, legislators, and the public were without much careful analysis useful in judging what became the leading proposals considered by Congress.

Getting Serious About Commodity Legislation, May–August 1995

2

Congressional attention to farm commodity legislation became increasingly more intense during the four months of May, June, July, and August 1995. The coming expiration of the 1990 farm commodity legislation was one reason for this intensity. Another reason, and perhaps the overriding reason for the increased attention to farm commodity legislation, was the "balance the budget" efforts, which moved into a new phase. In turn, congressional procedures for divining a balanced budget plan quickly dominated farm policy deliberations.

In the House, Congressman Pat Roberts, R-Kans., as chairman of the House Agriculture Committee, was forced to walk a tightrope between the House leadership and farm commodity groups, the mainstay of his political support. During this season he emerged as a cooperator on budget matters with the House leadership, but in the end even more as a shrewd defender and promoter of income transfers to commodity producers and farmland owners.

In the Senate, Sen. Richard G. Lugar, R-Ind., continued to embrace budget constraints and the scaling back of farm program outlays. However, he was unable to attract substantial support for his specific suggestion to phase out farm commodity programs by notching down commodity target prices by 3 percent each year. He chose not to force a committee vote on his proposal. Instead, he accepted the consensus that formed among his Republican Senate colleagues for a proposal of Sen. Thad Cochran's, R-Miss. Cochran's plan achieved estimated savings of federal outlays by reducing the acreage on which payments would be made to producers and owners of farmland.

During this season the administration, with great restraint, ultimately entered the farm bill arena. In the meantime the state governors and the House leadership were attempting to gain approval to block grant food stamp and school lunch programs as part of welfare reform. The Agriculture Committees in both the Senate and the House resisted these efforts.

New Balanced Budget Phase

The drive for a balanced budget plan moved into a new phase in May 1995. It was time for Congress to "talk their balanced budget talk." It was an ambitious undertaking. Specific seven-year (FY1996–FY2002) outlay numbers for entitlement programs (including farm commodity programs) and for activities supported with appropriations, like USDA research, export subsidies, and program administration, were to be agreed upon by Congress. Together these numbers were to constitute a plan for balancing the federal budget by FY2002. These balanced budget efforts were important to the farm bill debate. Not only was the Republican congressional leadership insisting that actual appropriations for FY1996 conform to the prospective plan, but they were also insisting that legislation like the prospective farm bill must conform to the balanced budget plan for all seven years.

A Game of Chicken

The interaction between the president and Congress over budget matters quickly became a game of chicken—who would identify specific programs for reduced outlays? The president had made the first move in February 1995 when he submitted his budget to Congress. It was not helpful to the Republican budget-cutters. His budget for FY1996, submitted in February 1995, called for only modest efforts to reduce the deficit. It appeared to reflect a strategic decision to force the Republican-controlled Congress to identify specific federal outlay reductions and program curtailments. The administration could then react to these specifics.

Given the Republican commitment to balance the federal budget, it became imperative for Sen. Pete V. Domenici, R-N.Mex., as chairman of the Senate Budget Committee, and Congressman John R. Kasich, R-Ohio, as chairman of the House Budget Committee, to make the first major moves in identifying the cuts that would be required to balance the budget by FY2002. And they did.

In May 1995 each chairman placed before his respective Budget Committee a plan for achieving a balance between receipts and outlays in the federal budget. They suggested maximum outlay numbers for FY1996–FY2002 for the many different outlay categories in the federal budget. These proposals were quickly considered and approved by the Republican majorities of the Budget Committees as resolutions. The Senate and House resolutions went to conference between the Senate and the House, and the resulting concurrent resolution on the budget was approved by both the Senate and the House.

The concurrent resolution on the budget approved by the Congress included instructions (limits on outlays) for those preparing the thirteen appropriations bills in Congress, including the FY1996 appropriations for the USDA, and authorizing committees, including the Senate and House Agriculture Committees, as they considered a new farm bill. With these outlay limits Congress placed constraints on any farm legislation arranged by the Agriculture Committees. These constraints were destined to have a major effect on the farm bill debate and ultimately had an overwhelming influence on the farm commodity policy chosen by Congress and accepted by the administration. During this period, budget considerations were so overwhelming that any discussion of alternative commodity policies started and ended with the question, "What budget estimates will CBO attach to the alternative?"

Role of Budget Resolutions

Two features of the procedures associated with congressional resolutions on the budget are particularly noteworthy, and awareness of them is important to an understanding of farm legislation activities in 1995. First, the provisions of budget resolutions constitute guidelines for congressional activities. For example, the resolutions limit the amount of money that can be appropriated by the appropriations committees and their subcommittees, like the Agriculture Appropriations Subcommittees headed by Senator Cochran in the Senate and by Congressman Joe Skeen, R-N.Mex., in the House. The resolutions also place constraints on policies approved by authorizing committees like the Senate and House Agriculture Committees that in the 104th Congress were headed by Senator Lugar and Congressman Roberts. For authorizing committees the constraints require the committees to find a combination of policies that generate CBO-estimated outlays no greater than those specified in the congressional resolution on the budget.

Second, the president is not required to respond to congressional budget resolutions. These resolutions simply are not submitted by Congress to the president since they are designed for possible self-discipline of Congress. Thus, in the budget chicken game, the president had the clear advantage. The Republican Congress had to be somewhat specific about proposed program cuts and risk adverse reactions by program constituents. The president did not have to respond. If he chose to do so, he could gain favor with those who might be hurt by proposed program cuts by criticizing the cuts without addressing the larger budget deficit problem.

Coping with Budget Resolutions—It's Easier for Some

For at least two reasons it was easier for Congressman Skeen's and Senator Cochran's Agriculture Appropriations Subcommittees to respond to the Budget Committees' instructions than for Congressman Roberts's and Senator Lugar's agriculture authorizing committees to do so. First, the Appropriations Subcommittees could essentially ignore (at least in 1996) the six future years because of the one-year cycle on appropriations, even though the concurrent resolution on the budget included related numbers for all seven years, FY1996–FY2002. The Agriculture Appropriations Subcommittees could concentrate on FY1996, test the "boundaries" between their appropriations activities and the agriculture authorizing committees, meet the FY1996 instructions of the Budget Committee, and let everyone assume that the agriculture appropriations in the other six years would be consistent with the May 1995 Budget Committees' instructions for those years, FY1997–FY2002.

But the agriculture authorizing committees could not follow the approach of the Agriculture Appropriations Subcommittees. Commodity legislation has traditionally been multiyear in duration. Since the Budget Committees were working with seven years, agriculture authorizing committee leaders were soon talking in terms of a seven-year farm bill. This approach meant that the farm bill proposals would be scored by CBO for each of the seven years FY1996–FY2002.

Second, CBO budget-scoring procedures for appropriations are fundamentally different than they are for entitlement-type legislation handled by the authorizing committees. CBO scoring of appropriations is straightforward; scores are merely the amount of money appropriated. Scoring of legislation like the farm bill is much more complex. Aside from the specific policy provisions included in the farm legislation, CBO estimates of actual federal outlays in any one year are based on a set of assumptions about several variables, including weather, anticipated market conditions, how programs will be administered, and how eligible participants will respond to program provisions.

In the Senate, once the concurrent resolution on the budget was approved by Congress, the Appropriations Committee, headed by Mark Hatfield, R-Oreg., gave an FY1996 allocation to its Agriculture Subcommittee. That subcommittee was headed by Senator Cochran. In the House, the Appropriations Committee, headed by Bob Livingston, R-La., gave an allocation for FY1996 to its Agriculture Subcommittee, which was headed by Congressman Joe Skeen.

These two Agriculture Appropriations Subcommittees focused on direct appropriations for various programs and activities of the USDA, including, for example, the Agricultural Research Service, the Natural Resources Conservation Service, the secretary's office, and many other USDA offices and agencies, as well as special projects that Congress directs the department to fund, such as this or that program, research center, or activity in some particular state.

The major steps for FY1996 appropriations included

—reporting out plans by the respective House and Senate Agriculture Appropriations Subcommittees for meeting the FY1996 allocations their subcommittees had been given by their overall committees,
—voting on these plans by the respective full Appropriations Committees,
—voting by the Senate and the House on the USDA appropriations bills sent to the floor by their Appropriations Committees,
—negotiating among Senate and House conferees on the differences between the Senate- and House-approved bills, and
—voting by the Senate and the House on the bill agreed to in conference.

In turn, the USDA appropriations bill for FY1996 was completed and forwarded to the president. In this particular case, it was signed by the president. Thus, USDA avoided being entrapped by the government shutdown at the end of 1995 as the president and Congress struggled to gain the political upper hand on the balanced budget issue.

In comparison to the immediate necessity for the Appropriations Committees to focus on FY1996, Senator Lugar's and Congressman Roberts's agriculture authorizing committees had to directly consider outlay estimates for a number of years consistent with multiyear farm programs. As indicated above, the Agriculture Committee leaders did not necessarily have to devise a seven-year farm program; however, once seven years became the mode for a balanced budget plan, Agriculture Committee leaders quickly focused on seven years as well.

One measure of the complexities associated with the scoring of farm programs is the priority given to interactions between the staffs of the CBO and the authorizing committees. Committee members desire minimum CBO outlay estimates for their proposals, and they do not want any surprises. These can arise not only from changes in CBO's expectations as to the effects of various policy provisions but also because changes in current commodity conditions can lead CBO analysts to change their expectations about future commodity market conditions. Changing market expectations, timing of the CBO scoring, and the role of CBO scores in the 1995 drive for a balanced budget plan account partially, as is illustrated in later chapters, for several decisions that eventually led to the enactment of the 1996 Farm Act.

In summary, in May 1995 the balanced budget efforts of the Congress moved into a new phase as the budget resolution was hammered together and the respective committees were given instructions. These instructions made the search for 1995 farm bill legislation more involved. The congressional leadership's commitment to crafting a balanced budget plan meant any legislation that was not meshed into a balanced budget plan simply would not get to the floor. Thus, the prospective farm bill had to have CBO scores consistent with the budget marks contained in the congressional budget resolution.

Pat Roberts's Dominance in the House

House of Representative farm-related activities during the summer of 1995 reflected the skill and tenacity of Congressman Roberts, the chairman of the House Agriculture Committee. Roberts argued strongly against the agriculture budget cuts ("marks") that were initially presented to him by the House Budget Committee. He persuaded the House leadership to reduce the size of the cuts and to "back load" the agreed-to seven-year cuts by applying the larger proportion of the cuts to the later years.

Also, Congressman Roberts went to the mat and won the turf battle with Congressman Skeen, the chairman of the Agriculture Appropriations Subcommittee. During the floor debate on USDA's appropriations applicable to the twelve months ending October 1, 1996 (FY1996), Congressman Roberts established his dominance over farm-related legislation. With only one relatively minor exception, House critics of the farm programs who proposed adjustments to the USDA appropriations bill were outvoted by substantial margins. The combination of these outcomes suggested that the political power of farm commodity interests and related agribusinesses remained strong and that Congressman Roberts was determined to protect the income transfers associated with the farm commodity programs.

Fighting the House Leadership

The interaction between Congressman Roberts and the House leadership over the budget marks for the Agriculture Committee illustrate his skill in disagreeing with leaders of his own party while retaining his committee chairmanship, protecting the turf of the House Agriculture Committee, mastering congressional infighting, and above all demonstrating his loyalty to his constituency of farm producers and landowners.

Congressman Roberts's success at dissent is illustrated by his interaction with the House leadership over the size of the budget cuts applicable to farm programs. For example, during the weekend of May 4, 1995, House Republicans held a retreat at a Leesburg, Va., conference center. This conference took place at the time of intense negotiations among Republicans about budget cuts. The situation was so tense that forty-one members of the Congress boycotted the first conference session because of the cuts in farm program budgets proposed by Congressman Kasich, chairman of the House Budget Committee. Kasich had talked in terms of cutting $16 billion over five years. However, by the time of the Leesburg conference he had lowered the number to $11.9 billion. At the conference, according to the *Washington Post,* Congressman Roberts pressed to make the cuts $6 billion rather than $11.9 billion. Congressman Kasich said, in effect, that he would accept the "6" if it were turned upside down. The weekend ended with a presumed agreement on $9 billion.[1]

Subsequent to the Leesburg meeting Congressman Roberts continued to press for a smaller cut and finally accepted $8 billion, but with a commitment by the House leadership that enhanced Congressman Roberts's negotiating position with his colleagues on the House Agriculture Committee. The leadership agreed to revisit the issue in three years if farmland values should happen to decline or if a few other unfavorable conditions befell producers and farmland owners. That way, if adverse conditions in the farm sector should happen to develop, Congressman Roberts would have a hook to press for more generous farm programs. In the meantime, he could use the leadership's agreement to a third-year review to help persuade his constituents to support whatever program he offered within the $8 billion constraint. In the end, all of the forty-one aggies who had boycotted the first session at the Leesburg retreat because of Congressman Kasich's proposals on the farm cuts voted for the Kasich budget resolution bill as it passed the House 238 to 193 on May 18, 1995.[2]

Fighting for Turf

Congressman Roberts was equally impressive when challenging Congressman Skeen, chairman of the Agriculture Appropriations Subcommittee.

The rivalry between Skeen's Agriculture Appropriations Subcommittee and Roberts's Agriculture Committee became sharply focused as USDA's appropriations for FY1996 were being considered. The rivalry was not new; however, the pressure, by at least some participants, to have a clear-cut division of responsibilities between the two groups was new. The resulting understanding established the primacy of the House Agriculture Committee and, in particular, Congressman Roberts, with respect to programs like those that support the prices of farm commodities.

The rivalry between the Agriculture Appropriations Subcommittee and the Agriculture Committee goes back many years. For example, when Congress initiated the CRP, the program was funded by the Commodity Credit Corporation (CCC). By structuring the financing of the CRP in this way the House Agriculture Committee prevented the Agriculture Appropriations Subcommittee from having an opportunity to consider the initial legislation even though the Agriculture Appropriations Subcommittee would at a later date be called upon to reimburse CCC outlays including those associated with the CRP. The alternative would have been to authorize the program and then look to the Agriculture Appropriations Subcommittee to appropriate the necessary funds to implement the program. But that approach would have required the cooperation of Congressman Jamie Whitten, D-Miss., the chairman of the Agriculture Appropriations Subcommittee, who did not support the initiation of the program. Thus, by funding the CRP through the CCC, the Agriculture Committee sidestepped the Appropriations Committee.

Another example of the intermittently tense relations between the Agriculture Appropriations Subcommittee and the Agriculture Committee relates to USDA's FY1994 and FY1995 appropriations bills. With these bills the Agriculture Appropriations Subcommittee prohibited the use of appropriated funds for administering the honey marketing loan payments and price supports for honey that the House Agriculture Committee had authorized. This prohibition on the use of appropriated funds meant that the honey program could not be implemented even though approved by Congress and, in particular, by the Agriculture Committees.

The intensity of the rivalry between the committees was heightened during the summer of 1995 by activities associated with the concurrent resolution on the budget for FY1996. The resolution called for the achievement of a balanced federal budget by the year 2002 and designated maximum specific dollar amounts for budget authority, budget outlay, new direct loan obligations, and new primary loan guarantee commitments for each functional area of the budget for each of the seven fiscal years 1996 through 2002. These designated amounts were then distributed among the subcommittees of the Appropriations Committee (for example, the Agriculture Appropriations Subcommittee) and the authorizing committees (for example, the House Agriculture Committee).

The budget amounts assigned to the respective committees were presumably related to the programs over which the respective committees have jurisdiction. For example, the amount assigned to the Agriculture Appropriations Subcommittee for FY1996 included amounts for programs associated with USDA administrative responsibilities. Other examples included the secretary's office, research programs, information, rural development, the Special Supplemental Nutrition Program for Women, Infants, and Children (WIC), and the administration of commodity programs.

In comparison, the budget amounts assigned to Roberts's Agriculture Committee were for programs like the Federal Crop Insurance Fund, CRP, CCC, Wetlands Reserve Program (WRP), food stamps, and child nutrition and other nutrition initiatives. The costs of the commodity price support programs such as deficiency payment and loan deficiency payments are "charged" to CCC. In turn, CCC funding is replenished periodically by an appropriation recommended by the Agriculture Appropriations Subcommittee.

Conflict erupted when Congressman Skeen's Agriculture Appropriations Subcommittee marked up and reported out an FY1996 USDA appropriations bill. The bill included selected cuts in activities that the Appropriations Committee typically handles. But this time it also included provisions which would have curtailed some of the programs usually handled by the House Agriculture Committee. The associated outlay reductions were claimed by the Agriculture Appropriations Subcommittee as part of their contributions to balancing the federal budget and fulfilling the instructions they had received from the House Appropriations Committee pursuant to the instructions that had been given to that committee by the House Budget Committee.

The approach of Congressman Skeen's Agriculture Appropriations Subcommittee was quickly attacked by Congressman Roberts. He successfully persuaded the House leadership that the Agriculture Appropriations Subcommittee must meet their assigned budget targets by changes limited to the programs that they usually handle. Reports circulated that Congressman Richard K. Armey, R-Tex., proclaimed that the Agriculture Appropriations Subcommittee could make any decisions it chose but that savings associated with programs typically handled by Congressman Roberts's Agriculture Committee would accrue to the "credit" of the House Agriculture Committee, not Congressman Skeen's Agriculture Appropriations Subcommittee. With these decisions the conflict over responsibility and credit for budget savings finally subsided. This battle again reflected the critical role of CBO scoring to the legislative process.

By the time the USDA appropriations reached the House floor on July 19, 1995, the details of the mandated settlement between the committees had been worked out. Congressman Skeen introduced House Resolution (H.R.) 1976, the appropriations bill, with these words: "...as the Committee on Appropriations, we have poached on the area of the authorizing committee, so we have decided to have a prenuptial agreement and divide this territory up and to get a property settlement."[3]

Congressman Skeen then proceeded to introduce en bloc amendments (to H.R. 1976 as reported by the Agriculture Appropriations Subcommittee) to reflect the "property settlement." There were two major concessions by the Agriculture Appropriations Subcommittee included in Skeen's amendments. One concession canceled the budget limits on the CRP, the WRP, and the Export Enhancement Program (EEP) that had been included in the appropriations bill reported out by the Agriculture Appropriations Subcommittee and actually formally approved by the House Appropriations Committee. The other concession was to cancel the prohibition of "certain disaster payments for livestock feed producers who refuse crop insurance." These two concessions involved earlier actions by the Agriculture Appropriations Subcommittee that Congressman Roberts obviously had considered beyond the jurisdiction of the subcommittee and intrusive to the responsibilities of the Agriculture Committee.

In turn, other amendments offered by Congressman Skeen were designed to save sufficient money so that the Agriculture Appropriations Subcommittee could still meet its instructions from the Budget Committee for FY1996. The amendments called for reducing salaries and expense accounts of the Consolidated Farm Service Agency (CFSA) by $17.5 million; eliminating the Great Plains Conservation Program and thereby saving $11 million; reducing the Section 502 direct housing loan level from $900 million to $500 million and increasing the guarantee loan program from $1.5 billion to $1.7 billion, thereby saving $83.6 million; eliminating the Rural Development Loan Fund to save $37.6 million; and reducing the Rural Development Performance Partnership Program for "rural utilities" by $127 million (from $562 to $435 million). Thus, at this point of the legislative process there was a direct tradeoff—reduced funding for rural development in order to preserve the funding of activities of more direct interest to commercial farmers—the CRP, the WRP, the EEP, and Disaster Feed Assistance.

Congressman Richard Durbin, D-Ill., the ranking minority member of the Agriculture Appropriations Subcommittee, argued against the en bloc amendments. He particularly objected to the proposed cuts in the "502 Housing Program." He also objected to making livestock producers who could have but did not buy crop insurance eligible for disaster feed payments. Congressman Durbin voted "no" but lost. The vote on the en bloc amendment was 240 yeas, 173 nays, and 21 not voting.[4]

Fending on the Floor with Critics of Farm Programs

Several additional amendments offered on the floor to USDA's FY1996 appropriations bill dealt with programs that over the years have been under the authority of the Agriculture Committee, Congressman Roberts's domain, not the Agriculture Appropriations Subcommittee. Roberts withstood the challenges with but one exception.

On the first day of floor discussion, July 19, Reps. Dan Miller, R-Fla., and Nita M. Lowey, D-N.Y., stated that they would not offer amendments regarding the sugar and peanut programs. Their willingness to do this was in exchange for a commitment by Congressman Roberts that the concerns about these programs would be considered by the Agriculture Committee. Both Representatives Miller and Lowey asked Congressman Roberts for assurances that "we will be afforded the opportunity to debate and vote on our amendment to the sugar program," in the case of Congressman Miller[5] and "that there will be an opportunity to discuss and vote on this issue [federal peanut program] on the floor during debate on the farm bill," in the case of Congresswoman Lowey.[6] Congressman Roberts responded "yes" to both and Representatives Miller and Lowey agreed not to offer their amendments.

Congressman Roberts surely knew, and Representatives Miller and Lowey should have known, that Roberts's response was likely a hollow gesture. At that time the expectation was still that key provisions related to commodity programs would be incorporated into the budget reconciliation bill under the jurisdiction of the Budget Committee. It was reasonable to anticipate that there might be opportunities to vote in the Agriculture Committee on particular commodity provisions as the committee responded to instructions from the Budget Committee. However, the probabilities were nil that there would be opportunities to offer amendments on specific farm commodity provisions buried in a giant budget reconciliation bill during its floor consideration in the House. Admittedly, legislators are adept at finding legislative vehicles to attach their proposals to when expected procedures are stifled. Further, from a tactical view, it is not clear why Congressman Roberts did not acquiesce to Representatives Miller and Lowey by allowing them to introduce their proposals on July 19 as amendments to the appropriations bill. The votes on the other amendments that were introduced suggest that Miller's and Lowey's amendments would have been easily defeated. Having had the votes on record would have made the Agriculture Committee's deliberations about commodity programs less complicated.

The votes on several amendments demonstrated the strength of the commodity and related agribusiness groups, as well as Congressman Roberts. Twelve of the amendments offered on the floor to USDA's FY1996 appropriations bill were of special importance to the commodity and agribusiness interests. The vote counts on eleven of the amendments favored these interests, most by large margins. The closest vote was the 196 yeas to 232 nays vote on the Sanford amendment to prohibit use of funds for a new office facility complex at the Beltsville Agricultural Research Center (see Table 2.1). The second closest was the 199 yeas to 223 nays vote on the Durbin amendment to prohibit the use of funds for extension work for tobacco and/or crop insurance for tobacco.

One amendment proposed eliminating from the bill a 7.3 million cap on the number of participants in the WIC program. It passed 278 to 145.

Other votes dealt especially with rural development programs. Two observations are particularly relevant to appraising the strength of commodity and agribusiness groups relative to other groups, like those who support rural development and rural housing. First, the amendments related to rural development programs largely dealt with shifting money among those programs. They did not call for shifting money from commodity to rural development programs. This reality, coupled with the manner in which the Agriculture Appropriations Subcommittee achieved compatibility with their budget instructions, suggests that informal understandings limited any possible shifts of money toward rural development and that commodity interests continued to have stronger influence than did the rural development interests.

Second, several of the rural development-related amendments focused on assistance for rural housing. Members of Congressman Roberts's Agriculture Committee claimed great allegiance to these programs, especially the Section 502 program, and suggested that they would do all possible to find money to replace the funds that the appropriations bill was taking from the programs. But this was largely posturing. The reality was that most of these same representatives were party to the decision to cut the housing programs rather than cut programs that supported commodity and agribusiness interests.

TABLE 2.1. House votes on selected proposed amendments to H.R. 1976, USDA Appropriations, July 1995

Sponsor (Roll no.)	Thrust of amendment	Voice	Roll Call Aye	Roll Call No	Not Voting	Present
Allard (539)	Cut funds for several offices in the higher echelons of the USDA by 10%		196	232	6	
Sanders (541)	Reduce MPP by $3 million and devote $2 million to BST related analysis and tests		70	357	7	
Durbin (544)	Prohibit use of funds for extension work for tobacco or tobacco crop insurance		199	223	12	
Zimmer (550)	Eliminate the proposed $110 million funding of MPP		154	261	19	
Lowey (545)	Prohibit deficiency and land diversion payments to persons with annual adjusted gross income of $100,000 or more from off-farm sources		158	249	19	8
Minge	Restrict Lowery's prohibition to those living in incorporated municipalities with a 1990 population of more than 50,000	no				
Hoke (547)	Reduce proposed PL 480 Title I appropriations by $113 million to the level requested by the President		83	338	13	
Sandford (548)	Prohibit use of funds for new office facility campus at Beltsville Agricultural Research Center		199	221	14	
Oliver (549)	Prohibit assistance to livestock producers if crop insurance protection or non insured crop disaster assistance for the loss of feed produced on the farm is available		169	248	17	
Obey (551)	Prohibit MPP assistance to organization with annual gross of $20 million or more unless it is a cooperative		176	229	29	
Kennedy (552)	Prohibit promotion of alcohol or alcoholic beverages with MPP		130	268	36	
Deutsch (553)	Prohibit assistance with MPP to U.S. Mink Export Development Council		232	160	42	
Hall (543)	Eliminate cap in bill of 7.3 million WIC participants		278	145	11	
H.R.1976	H.R. 1976 as amended		313	78	43	

Source: *Congressional Record*, 104th Cong., 1st sess., July 19, 20, and 21, 1995, 141, pt. 83-85.

The Anonymous Three-Pager in the House

During July 1995 a three-page paper began circulating among congressional offices interested in commodity policy. Seldom has a paper moved so quickly among the farm commodity and agribusiness lobbyists in Washington, D.C. The title of the three-pager was "The Freedom to Farm Act of 1995." Its subtitle was "A Seven Year Contract with Production Agriculture." It was anonymous and not dated (see Appendix 3).[7]

Seven-Year Annual Payments

The paper called for making annual payments to farmers over seven years. Those who had received payments from the CCC in three of the past five years would be eligible to sign contracts. The government would agree to make the payments and farmers would agree "in exchange" to continue conservation practices that had been previously developed.

Government programs designed to adjust acreage of crops from year to year in response to price, supply, and demand conditions would be discontinued.

Outlay estimates for the proposed program were not included. However, according to the paper, total dollar outlays for each of the seven years, 1996–2002, would be fixed, thus responding to one of Senator Lugar's criticisms of past commodity programs—the inability to reliably anticipate farm program outlays.

The paper did not mention anything about what would happen after 2002, the seventh year of annual payments.

The Welfare Ghost

In a manner somewhat similar to many staff papers, this three-pager included a list of advantages and disadvantages. The advantages were the usual arguments for free markets. One identified disadvantage suggested that the proposed "Freedom to Farm Act" would be essentially a "capped entitlement" and the proposal would "move farm program payments into the category of 'welfare payments.'"

However, this welfare ghost—the possibility that the proposed "capped entitlement" would be viewed by the public as another welfare program by farm or nonfarm groups—never materialized. This may have been because neither the three-pager nor the eventual proponents of this approach suggested that the proposed payments be targeted to low-income producers or that the payments be inversely linked to income levels of recipients with low-income recipients receiving larger payments than higher-income recipients. Nor was it suggested that certain levels of owned assets would disqualify producers or landowners from receiving the proposed payments. The sole criteria proposed for being eligible for the payments was whether commodity farm program payments had been received in three of the past five years. No current programs commonly labeled as welfare programs have rules where having received benefits in the past is prima facie qualification for receiving benefits for even one more year into the future, much less seven years into the future. Thus, in fact, the proposal had little in common with the restrictions that most people associate with welfare programs. In turn, the public did not make the comparison feared in the three-pager.

Potential Turf Issue

On a more aggregative level, the suggested program approach would have budgetary characteristics similar to the programs for which the Appropriations Committees presently have jurisdiction. The amount of outlays for any one year would be fixed and could be very closely anticipated. Committee jurisdiction on this matter was not an issue during 1995, nor has it become an issue since. However, if the program were to be extended beyond 2002, it could be.

Congressman Roberts's Wariness and Then Embracing of Freedom to Farm

Congressman Roberts carefully and cautiously moved toward publicly supporting the concepts included in the three-pager.

Step by Step

In mid-July 1995, news sources reported that Congressman Roberts was preparing a bill that would end the idling of cropland and the necessity of farmers to plant particular crops in order to qualify to receive price-support loans and associated income transfers. In describing the program to news people, Roberts focused on year-to-year payment variability of past farm programs and ascribed frustration to farmers not knowing what the size of their government payments would be in any one season.

Within a week, however, Roberts indicated to reporters that he had not endorsed any approach to the farm bill. Even so, he suggested that the Freedom to Farm legislation being drafted by the Agriculture Committee staff merited serious study.[8]

On August 4 Congressman Roberts introduced H.R. 2195, the Freedom to Farm Act of 1995. Congressman Bill Barrett, R-Nebr., chairman of the House Committee on Agriculture, Subcommittee on General Farm Commodities, was a cosponsor. Thus, the debate over prospective commodity policy, invigorated by the anonymous three-pager, was energized just as Congress started its August recess. The stage was being arranged for the markup of Roberts's proposed farm legislation by the House Agriculture Committee when its members returned to Washington, D.C., in September.

What to Stabilize

During the summer of 1995, the traditional rhetoric about farmers' needs for stability was flipped inside out by Congressman Roberts.

Over many years agricultural leaders argued that given the instability of commodity prices and yields, government payments should vary from year to year in order to stabilize farm income—revenue from the market plus farm program checks. But now Congressman Roberts, the most powerful aggie in Congress in terms of farm programs, was implicitly arguing for instability of income. Most everyone understood that a combination of stable farm commodity payments coupled with unstable market prices meant unstable income.

Two factors tended to provide an acceptable degree of credibility to Congressman Roberts's proposition that farmers should know early on the size of their government farm check. The first related to commodity market price expectations. In the summer of 1995 there developed a fairy-tale dimension to the prevailing attitudes toward future farm commodity prices. Market prospects were bullish. U.S. wheat prices were higher than at any time earlier in the 1990s. Corn prices were not, but they were fifty cents per bushel above lows reached in late 1995 and forecasts were for them to go higher. Cotton prices were a full ten cents per pound off their early 1995 high, but still above the highs of 1991 and early 1994. It was easy to dream that exports were moving farm commodity prices to new, higher, and long-term sustainable plateaus and to confuse the dream for reality. Anticipation of greatly increased exports to China was reflected in the enthusiastic optimism.

The second factor that gave Congressman Roberts's proposal credibility is the tendency to confuse the concept of stability with the concept of decline. It is not so much price instability that sellers of products dislike as it is price declines. They like price increases. Advocating fixed farm commodity program payments resonated favorably with those who believed that high commodity price plateaus were the new order. It was a dream combination. Ride what is going up (commodity prices) and put a floor under what under the then current farm programs would go down (farm payments). Not only that—also convince the leadership of the Congress, as Congressman Roberts had, that if things didn't work out, that is if farm commodity prices dropped and land values crashed, the decision would be revisited.

Searching for Consensus in the Senate 3

In the Senate, Sen. Richard G. Lugar, R-Ind., openly supported scaling back the outlays for commodity programs. However, he was unable to garner support for his proposals even among his Republican colleagues. The Democrats appeared not as sharply divided as the Republicans. They seemed united in wanting to continue commodity programs into the foreseeable future, but they had difficulty in forming a consensus around a specific approach that had wide appeal.

Aside from the inability to develop a consensus on the future direction of commodity programs, the most notable development among Senate aggies during the summer of 1995 was the largely unnoticed cooperation of Senator Lugar's and Sen. Patrick J. Leahy's, D-Vt., staffs. They were able to jointly draft and agree on conservation and research titles for insertion into a farm bill when the time became propitious. This bipartisan cooperation reflected the work of astute staff members for each senator. This cooperation, although somewhat modest, nonetheless was in sharp contrast to the situation among members and their staffs of the House Agriculture Committee.

In other respects the situation in the Senate was not much different than in the House. Budget considerations had an overwhelming influence on the deliberations. Also, most aggies, Republicans as well as Democrats, favored continuing farm commodity programs much as these programs had operated under the 1990 Farm Act.

Senator Lugar's Approach

In the Senate the chairman of the Agriculture Committee, Senator Lugar, approached farm commodity legislation in a dramatically different way than did Congressman Pat Roberts, R-Kans., in the House. While Congressman Roberts struggled mightily with the chairman of the House

Budget Committee to minimize budget cuts in farm commodity programs, Senator Lugar embraced proposed cuts in farm program outlays and even proposed deeper cuts than those suggested by the chairman of the Senate Budget Committee, Sen. Pete V. Domenici, R-N.Mex.

Congressman Roberts was circumspect until August 1995 as to which approach to commodity policy he favored. In contrast, Senator Lugar attempted to initiate a broad debate on government's role in agriculture and was quite specific early in 1995 about his proposal—phase out the programs by cutting target prices by 3 percent per year. The lowering of the target prices would decrease the amount of deficiency payments paid by the federal government to producers and landowners (relative to what they would be with a continuation of the 1990 Farm Act).

Congressman Roberts introduced legislation incorporating the approach he came to embrace; Senator Lugar never did. Both individuals, however, embraced the notion that government involvement in production agriculture should be relaxed. In some ways, Senator Lugar's approach was potentially more draconian for producers than Congressman Roberts's. It would have brought an end to farm commodity deficiency payments. In contrast Congressman Roberts's approach was very ambiguous as to what might happen after 2002.

The reality was that Senator Lugar's target price-cutting approach was not acceptable to his peers. Of course, he could have chosen to introduce legislation during the summer of 1995 (or later in the fall) that included his plan for decreasing target prices by 3 percent per year, but he did not. The best guess is that if he had, the bill might have attracted support from only one other member of the Senate Agriculture Committee, Sen. Rick Santorum, R-Pa.[1]

Debating the Democrats

Democrats on the Senate Agriculture Committee were very critical of the Republicans' drive to develop a balanced budget plan. They were particularly critical of Senator Lugar's suggested approach to achieve budget savings by cutting farm commodity target prices. For example, Sen. Kent Conrad, D-N.Dak., a member of the Senate's Budget Committee, directly challenged Senator Lugar at a February 16, 1995, meeting of the Budget Committee. Senator Lugar, who was not a member of the Budget Committee, testified before the committee and repeated his proposal for cutting farm commodity target prices. In turn, Senator Conrad, a member of the Agriculture Committee as well as the Budget Committee, directly challenged Senator Lugar with a presentation of his own. He argued that if there were no farm program, wheat prices would drop drastically and in North Dakota bankruptcies would be widespread.

Sen. Byron L. Dorgan, D-N.Dak., although not on the Senate Agriculture Committee, spoke strongly in support of commodity programs. He advocated revising the commodity programs so that the program benefits would be more concentrated with smaller rather than larger farms. In an opinion piece distributed in connection with an April 21, 1995, forum in North Dakota, Senator Dorgan emphasized concepts that he thought should be in a farm bill—concepts to which he would return many times over the following ten months. He proposed that the target price for wheat, for example, be increased from the then current $4.00 per bushel to $4.50, but that the deficiency payments based on the difference between market prices and the target prices be available on only the first twenty thousand bushels produced by a farmer. Similarly, he advocated that nonrecourse loans be available for only a limited number of bushels for any one producer. Senator Dorgan made these specific suggestions in the context of arguing that the programs should be structured to support small or moderate-sized family farms rather than designed to channel the major portion of federal funds to large farms.[2]

Senator Dorgan often repeated his appeal to change the farm commodity programs in ways that moved away from equal program benefits per unit of output, which tilted benefits to larger producers, towards greater equality of benefits per producer. The concept was included in a statement of Sen. Thomas A. Daschle, D-S.Dak., in August 1995[3] and more formally incorporated in S. 1256, introduced in the Senate on September 18, 1995. Cosponsors included Sens. Patrick J. Leahy, D-Vt.; Byron L. Dorgan, D-N.Dak.; Kent Conrad, D-N.Dak.; J. Robert Kerrey, D-Nebr.; Tom Harkin, D-Iowa; Paul D. Wellstone, D-Minn.; James Exon, D-Nebr.; Max Baucus, D-Mont.; and Wendell H. Ford, D-Ky. However, Senator Dorgan's approach was generally ignored by leaders of producer groups, the administration, and the legislators who in the end had the major influence on the making of the 1996 Farm Act.

THE AGRICULTURE DEFENDERS

For the most part, Republican and Democratic senators with major interest in commodity legislation defended the past programs and were anxious to protect the programs from budget cuts. For example, Sen. Charles Grassley, R-Iowa, who was a member of both the Senate Budget Committee and the Senate Agriculture Committee, demonstrated his allegiance to commodity programs and especially their funding by gaining approval of a nonbinding resolution in the congressional concurrent resolution. It called for 80 percent of the budget reductions assigned by the Agriculture Committee to be applied to nutrition programs and 20 percent to agricultural programs.

However, not all senators who were closely associated with past commodity programs were resolute in their attachments. For example, as the summer progressed Senator Leahy, ranking member of the Senate Agriculture Committee, became increasingly concerned about the challenges to food and nutrition programs that were being mounted and supported by farm commodity interests. In a speech on the Senate floor in mid-July, Senator Leahy vented his frustration over the attacks on food and nutrition programs and indicated that his support of commodity programs was conditional. He stated, "If farm programs become the enemy of the hungry and the environment, I will not support them. Indeed, I will join those on the floor who want to dismantle them."[4] He may have been talking as much to his Democratic colleagues as to the Republicans.

Among the commodity groups, cotton and rice groups were increasingly concerned that Congress would change policies to be like the proposal described in the three-page "freedom to farm" paper. Although they favored less regulation, they nonetheless also favored benefits geared to market prices as provided under the 1990 Farm Act and before. In response to their attitudes Sen. Thad Cochran, R-Miss., introduced S. 1155 on August 10, 1995. The bill would leave the major mechanisms of commodity programs essentially unchanged, but nevertheless generate substantial savings. The savings would be obtained by lowering the percentage of crop acreage eligible for deficiency payments from 85 percent to 75. Cosponsors included Sens. David Pryor, D-Ark.; Paul Coverdell, R-Ga.; Jesse Helms, R-N.C.; John W. Warner, R-Va.; Larry E. Craig, R-Idaho; Sam Nunn, D-Ga.; Trent Lott, R-Miss.; J. Bennett Johnston, D-La.; John B. Breaux, D-La.; Strom Thurmond, R-S.C.; Connie Mack, R-Fla.; Daniel K. Inouye, D-Hawaii; Daniel K. Akaka, D-Hawaii; Dale Bumpers, D-Ark.; and Mitch McConnell, R-Ky.

Other Summer Activities

In the Senate, the chairman of the Agriculture Appropriations Subcommittee, Senator Cochran, deferred consideration of USDA's FY1996 appropriations until the shape of prospective commodity legislation was better known. In addition, without fanfare Senator Lugar and his staff worked carefully with Senator Leahy and his staff to craft research and conservation titles for use when the opportunity to include them in farm legislation arose later on. In the meantime Senator Lugar defied other Senate Republicans who were attempting to block grant the food stamp program as part of welfare reform.

Thus, as the summer of 1995 ended most senators with farm interests, Republican and Democrat alike, were in a defensive mode. Senator Cochran introduced legislation that would continue the then current program but with adjustments downward in the acreage for which the producers would receive payments. Senator Lugar did not introduce legislation. He was unable to persuade other Republicans or Democrats on the Senate Agriculture Committee to support what he favored—a lowering of commodity target prices by 3 percent per year. Nonetheless, he persisted in favoring his proposed phaseout of commodity programs.

The Democratic senators from the Dakotas wanted the commodity programs to continue, but repeatedly pressed for a change in the distribution of the benefits. They argued that the amount of benefits to any one recipient should be limited. Research and conservation were areas in which Senators Lugar and Leahy were able to cooperate, and they did so by having their staffs jointly develop legislative titles. These titles developed by cooperating staff members were destined to become key parts of a bargain made in 1996 that sealed the outcome of the search for a farm legislation to replace the 1990 Farm Act. And, with a hint of what was to happen in February 1996, Senator Leahy reminded his colleagues in July 1995 that his priorities were food, nutrition, and environment. His support for farm commodity programs was not immutable but rather contingent on them being part of a larger framework.

The Administration's Restrained Entry

4

In May 1995 the administration timidly entered the farm bill arena with the release of its USDA-prepared "Blue Book." Also, during the summer newly appointed Secretary of Agriculture Dan Glickman made a limited number of statements about prospective commodity legislation; the president did as well. Any administration proposals for changes in commodity programs were quite modest, however. The controversy surrounding Secretary Espy's departure and the appointment of a new secretary took energy that might have otherwise been devoted to the farm bill process.

The administration's cautious approach was consistent with an expectation that the Republicans might propose substantial cuts in farm programs as part of their efforts to balance the budget. If they did, the administration would have an opportunity to challenge the proposed cuts.

In addition, past congressional attitudes toward assertive innovations on commodity policy by both Democratic and Republican administrations suggested that there was little for the administration to gain by being forceful in proposing changes in commodity programs and maybe a lot to lose. "Let the Republicans take the heat for any substantial contraction of the benefits associated with farm commodity programs" seemed to be a key part of the administration's approach to farm policy.

The administration's approach was evident when administration officials aggressively responded to the Republicans' proposed budget levels for food programs. They fueled the heat by portraying the Republicans' proposed House budget numbers for future years as representing cuts from current levels when, in fact, the Republican-proposed numbers represented increases from current program levels but less than were projected with no change in policy. This game of confusing the public over what was meant by budget "cuts" was rife throughout 1995 as the administration and Congress struggled over the balanced budget issue.

The Blue Book

With the prospect that members of the congressional Agriculture Committees would be critical of the administration actively advocating commodity policy, the administration followed the practice of earlier administrations. The USDA prepared and released (after being cleared by White House offices) a document that mixed together descriptions, opportunities, and modest prescriptions. This time the document was referred to as the "Blue Book." It was a ninety-four-page publication entitled *1995 Farm Bill: Guidance of the Administration*.[1] The central message of the document was to keep the current programs with some adjustments that would be relatively easy to accomplish and would generate some, but modest, savings of federal outlays. It was very cautious about change.

The Blue Book's guidance regarding planting flexibility illustrates the administration's caution. The book described how then current programs affected cropping practices of farmers. It cited the flex acre feature of commodity programs and identified the "normal" flex acres program provision whereby producers and farmland owners did not receive payments on 15 percent of their base acres of particular commodities including wheat, feed grains, cotton, and rice. Also, the "optional" flex acres program feature (up to an additional 10 percent of base acres) for which producers lost payments if these acres were planted to an alternative crop was mentioned. The Blue Book allowed that, "An effective way to achieve greater planting flexibility is to combine all crop bases into a total acreage base (TAB)." Then it suggested that, "The percentage of the TAB that producers could plant with alternative crops designated by the Secretary would be gradually increased from the current level of 15 percent to possibly 100 percent over 5 years." And, in response to the rather widespread concerns among agricultural groups that the department had cut back crop acreages too often and by too much, it stated that withholding land from production with the Acreage Reduction Programs should become "a discretionary tool" and used only "when market supply and demand are critically out of balance."

Examples of other proposals which were advocated in the Blue Book or placed in a "should be considered" context included

—Limiting target program payments by allowing payments only to individuals who earn less than one hundred thousand dollars annually from off-farm sources.
—Developing a program whereby producers in environmentally sensitive areas are guaranteed commodity program payments if they

operate their farm according to a "whole farm conservation plan."
—Developing a pilot program to encourage producers to save in high income years.

The pilot program example given in the Blue Book called for the federal government to match contributions by producers to special savings accounts up to an unspecified percentage of the producer's income. The accumulated money in these accounts was to be used when incomes were low. The proposal had similarities to Canadian programs designed to even out farm revenue through income transfers to producers. It reflected the notion that the federal government has a responsibility to not only supplement but also to stabilize the income of producers, and then a responsibility to encourage them to save in high-income years. The suggestion illustrates the special place transfers to agricultural producers and farmland owners have in U.S. politics. One wonders how many other groups, including welfare recipients, would be willing to increase their savings if the federal government matched their contributions to a savings account.

Sen. Richard G. Lugar's, R-Ind., comment to reporters in May 1995 regarding the Blue Book is indicative of the attention the book received from members of the Agriculture Committees. He told reporters that the proposals included in the Blue Book were "largely irrelevant" and repeated his support for cutting target prices by 3 percent per year and eliminating the Export Enhancement Program.[2]

Leave It to the Republicans

As the summer progressed it was clear that the administration's strategy was to let the congressional Republicans divine an approach to farm commodity programs compatible with Congress's own drive to balance the federal budget. There was no political advantage for the administration to propose significant cuts in farm commodity programs. Furthermore, this approach to commodity programs was consistent with the administration's approach to the overall "balance the federal budget" effort. Secretary Glickman's comment about budget cuts when he released the Blue Book on May 11 illustrates the "leave it to the Republicans" strategy. On that occasion he suggested that any cuts greater than $1.5 billion over five years would be "extremely unwise."[3] The $1.5 billion was equal to the claimed savings for the changes included in the Blue Book and was presumably relative to likely outlays if the programs were continued unchanged, but that relationship was not made clear.[4]

Although the administration's strategy of letting Congress make the first moves on commodity legislation was consistent with the administration's overall balanced budget strategy, it may not have been the only reason to let Congress take the lead. Congress, over the years, has treated with disdain most major proposals to revise farm commodity programs put forth by administrations—Republican and Democratic alike. This reality can cause any secretary of agriculture (and president, for that matter—if they happen to be interested in farm policy) to be very cautious about aggressively advocating major changes in farm policy.

Regardless, the administration's cautious approach to changes in commodity programs continued throughout the summer months. For example:

—Secretary Glickman in May 1995 when discussing possible changes in farm programs suggested that the proportion of base acreages for which producers and landowners do not receive government payments could be increased from 15 percent to 17 percent.[5]
—On another occasion in May when pressed as to how the administration would propose to enforce their suggested ban on farm program payments to those who receive one hundred thousand dollars from off-farm income, the secretary declared that the administration would let individuals declare if they were ineligible under the proposed rule.[6]
—In a possible reference to the agreement that Congressman Pat Roberts, R-Kans., had reportedly worked out with Congressman John R. Kasich, R-Ohio, for a $9 billion savings in farm programs over seven years, President Clinton when visiting Montana in June stated that cuts of $8 billion to $9 billion in farm programs over five years were "excessively large."[7]
—In June, Secretary Glickman told reporters previous to a Senate Agriculture Committee hearing (at which he was to outline ways to save money) that he was inclined to favor an increase in flex acres—make payments on fewer crop acres. But, he was not specific.[8]

Also in June, Sen. Patrick J. Leahy, D-Vt., indicated that the administration's new budget plan would propose a $4 billion cut in farm programs over seven years. This amount compares to the initial administration number of $1.5 billion of cuts in farm programs over five years.[9]

One of the more relatively audacious proposals by the administration was made by Secretary Glickman in the latter part of June. He proposed combining wheat, feed grain, cotton, and rice bases into a total acreage base for individual farms with operators having the freedom to change the mix of seven crops. They could also plant other crops on up to 15 percent of the total acreage base. This percentage would rise over time.[10]

Quarreling Over Food Programs

Food programs have had spillover political effects on farm commodity programs. For many years congressional observers noted that support for food programs encouraged reciprocal support for commodity programs and vice versa in the Agriculture Committees and on the floor of the House and Senate. Thus, approaches to welfare reform which encompassed food programs engendered concerns by commodity interests. Most importantly, budget procedures placed food and commodity programs in direct competition for budget allowances. Also, during the summer of 1995 the potential political payoff of "let the Republicans take the heat for budget cuts" was illustrated by an episode associated with Republican proposals on food programs as they worked on welfare reform. The administration's criticism of the Republicans' approach to food programs was enhanced by the public's confusion as to whether budget cuts were from past levels of program outlays or from projected levels (usually higher) based on current policies and programs.

The Administration

During the summer months the administration took opportunities to demonstrate support for food programs like school lunches and food stamps, but they also signaled some willingness to restrain future increases in outlays for these programs. A White House fact sheet released in mid-June stated that, "For food assistance, the President would maintain the national nutrition safety net programs while cutting mandatory (food program) spending by $20 billion over seven years."[11]

Also in June, Secretary Glickman, when attending a policy forum, commented about the linkages between food and farm programs. He stated that both farm programs and nutrition programs would be in "deep trouble" if the link between the two were not maintained. The press also reported that the secretary contended, "Eliminating farm programs would raise food prices and put even more pressure on poor families and government nutrition programs." Further, he claimed that, according to the media, "the coalition between farmers and supporters of nutrition programs was responsible for keeping food prices steady in real terms for half a century."[12]

Two Types of Concerns

The effect of the drive to balance the budget on food programs (for example, food stamps) was of concern to those interested in commodity legislation. These concerns were of at least two types.

First, although the Budget Committees' deliberations focused specifically on food programs, their instructions to the Senate and the House Agriculture Committees specified that the authorizing committees were required to meet the overall budget numbers given to them for *all* of the programs under their jurisdiction. The notion was that so long as they met the overall number, they could allocate the budget numbers among the programs for which they had authority. This approach placed the final responsibility on the authorizing committees to decide on the allocations between commodity programs and food programs. Thus, so long as the Agriculture Committees were unable to negotiate relief from the overall budget number given to them by the Budget Committees, the commodity programs and the food programs were in direct budget competition with each other. Also, within the commodity programs, commodities were in budget competition with each other.

Second, the past practice of logrolling votes between food programs and commodity programs was threatened by the Republican approach to welfare. Many Republican senators, members of Congress, and Republican governors wanted to block grant programs to the states as part of welfare reform and as a way to generate estimated budget savings. If block granted, the food programs represented an attractive amount of money over which the governors would have some control, including the possibility of shifting money from food programs to other welfare programs for lower-income people in their states.

Part of the concern among congressional aggies was that the decision to block grant the food stamp program, for example, would entail decisions by several individuals who were not members of the Agriculture Committees. Thus, it would be more difficult for the Agriculture Committees to work their will in any one year on the amount of money available for blocked food programs. Those members of Congress interested in food programs would be freer to vote their will on commodity programs. In instances when they chose to vote against bills supported by the commercial commodity and agribusiness interests, there would be less chance of being punished by Agriculture Committee actions on food legislation. Similarly, those interested in commodity programs would be freer to vote their will on food programs with less chance of retribution by those favoring food programs.

In the Senate

Both the Senate and the House were indecisive on whether to move ahead with block granting the food stamp program. In the Senate, the Committee on Finance had the lead responsibility to develop the Senate's approach to welfare. In April 1995 its chairman, Sen. Robert Packwood, R-Oreg., indicated that he favored block granting Aid to Families with Dependent Children (AFDC), as well as food stamps to the states.[13]

Although some senators from farm states were willing to consider the possible block granting of the food stamp program, they steadfastly favored keeping federal control of the programs for children like the school lunch program. Senator Lugar, for example, was quoted in May as being inclined to be more "bold" than other senators on food stamps but observed that, "I am more inclined to retain child nutrition programs at the federal level, while targeting benefits more."[14]

In contrast, Senator Leahy was opposed to block granting food stamps as well as other federal food programs and warned that he might try to apply the block grant idea to farm subsidies if nutrition programs were targeted for block granting. "If it works for one, it ought to work for the other," he stated.[15]

Other senators also pressed to keep the school lunch program as a federal program. Since the program is extremely popular with benefits flowing to middle-income as well as low-income families, the agriculture legislators naturally wanted to retain jurisdiction over the program. In arguing for keeping the school lunch program as a federal program, Sen. Thad Cochran, R-Miss., pointed out that "sixty percent of the public school students in my state get free and reduced price lunches." It is no accident that many photo opportunities for members of the Agriculture Committees include school lunch program settings.

Senator Lugar became convinced that block granting of food programs should not happen. In turn, in the first part of June he stated that he was opposed to block granting of food stamps as well as school lunch programs. When explaining his rationale for keeping the food stamp program as a federal program, Senator Lugar observed that states need extra money during recessions, and block grants would not automatically expand. In contrast, food stamp benefits as a whole adjust to changes in economic conditions like unemployment changes. AFDC families become eligible for coupons if their income falls to or below $2,027 per month. Also, if recipients are in states with lower AFDC payments, they get larger food stamp benefits.[16] These adjustments were seen as less likely to occur if states took over the program.

At the same time other senators supported block granting of food stamps thinking that if that were done the cuts in the budget for food stamps could be greater and therefore the cuts in commodity programs could be less.

Although the Senate Republicans decided during the summer not to block grant food stamps,[17] the quarrel over whether to block grant the programs or not was to continue well into the winter. Throughout the fall Senator Lugar, as a member of a House-Senate committee on welfare reform, steadfastly refused to agree to block granting of food programs. He thereby had a major effect on the slowing of the Republicans' effort to develop welfare reform legislation during that time period.

In the House

Congressional discussions about how to save money on entitlement programs included the possibility of block granting the food programs to the states along with other programs such as Medicaid. The Republicans in the Congress were not of one mind on this question. However, at least the more vocal Republican governors strongly encouraged block granting of food stamps. They saw the possibility of such block grants being an important source of money for other programs.

Actions by House Republicans on food programs offered the administration a unique opportunity to portray the House Republicans as mean-spirited even against schoolchildren. Partially because of Congressman Roberts's opposition to block granting the food stamp program, the Republicans decided to keep food stamps as a federal program, but offered to block grant the school lunch programs to the states. Coupled with the decision to block grant the school lunch program was a decision that the budget for the school lunch program over the 1996–2002 seven-year period should be less than outlays would be if policies were unchanged. These amounts for future years, although greater than in recent years, were less than the amounts the CBO estimated would be the outlays if current policies were extended into the future. The Republicans were taking steps to restrain the growth.

The administration chose to portray the approved budget numbers as "cuts" in the programs. The implication taken by the public was that the Republicans were cutting program outlays to levels less than in recent years. The Republicans were immediately placed on the defensive and were never able to surmount the semantic confusion between "cutting from a projected increase" and "cutting to a level below recent levels."

Part of the Republicans' difficulty was of their own making. In their enthusiasm for displaying their success in working toward a balanced budget, they emphasized cutting programs. Then when initially publicizing their decision, evidently it didn't seem important to them to emphasize that they were approving increases in the programs, but increases that were less than what CBO said the outlays would be if the programs were not changed. As a result the administration seized the opportunity to characterize the changes as program cuts. These kinds of nuances can be well understood by congressional staffers, as well as members. However, the differences can be easily lost when communicating to the public. And, in this instance, the administration was very successful in putting their spin on the developments.

As it turned out, in a manner consistent with the administration's overall approach to the Republicans' balanced budget efforts, the administration's activities related to the prospective farm legislation were very restrained. However, the Republicans took action on related food programs that offered the administration an opportunity to portray the Republicans in Congress as mean-spirited.

THE STATE OF PLAY AS CONGRESS RECESSED IN AUGUST 1995

5

The pressure on legislators to make specific commodity legislative proposals increased in late July and early August 1995. It was commonly accepted that the 1990 Farm Act expired at the end of 1995; the planting season for the 1996 winter wheat crop was near at hand, and it was expected that new legislation would be in place for producers to sign up for programs. However, the Budget Committees' insistence that the respective committees respond by September 22 on ways to reduce budget outlays (from anticipated outlays with unchanged policies) was undoubtedly the greatest impetus to the making of a new farm bill.

KEY AUGUST 1995 DEVELOPMENTS

In response to the pressures that peaked within a very short time period, there were several developments related to the prospective farm commodity legislation. These included

—On the first day of August Secretary Glickman released a state-by-state tally which purported to contrast what farmers would receive in federal payments if the Republicans' budget plans were adopted with what they would receive with the administration's plans.[1]
—Congressman Dick Zimmer, R-N.J., who on July 11, 1995, had introduced H.R. 2010, expressed interest in Congressman Pat Roberts's, R-Kans., Freedom to Farm act. H.R. 2010, if adopted, would have phased out commodity support programs over six years and capped total farm commodity payments during the phaseout years. However, Congressman Zimmer noted that the draft legislation did not call for a phaseout of the programs at any time.[2]
—Congressman Steve Gunderson, R-Wis., on August 3 made public a proposal to reform U.S. dairy policy and on August 17 intro-

duced a bill calling for eliminating the current assessment on milk producers who increase production, changing the milk marketing order system so that there would be only five or six orders, replacing the then current milk pricing system, and initiating a recourse loan program for processors of dairy products.[3]

—The administration acknowledged the budget implications of the bullish prices for most farm crop commodities. Everyone who cared had recognized the implications; however, these budget implications had not been openly acknowledged in the political debate over farm policy. Commodity markets were much stronger in late 1994 and early 1995 than had been anticipated. Although July 1995 cotton prices were lower than December 1994 prices, the opposite was the situation for other key commodities supported by government programs. Wheat prices were up by thirty-six cents per bushel, corn prices were up by fifty cents per bushel, soybean prices were up by fifty-nine cents per bushel, and rice prices were up by sixty-nine cents per hundredweight as reported by USDA's National Agricultural Statistics Service. These price increases led analysts to adjust their expectations of future prices upward. Thus, prospective federal outlays (payments to producers and farmland owners) associated with a continuation of current commodity policies would be substantially lower than both USDA and CBO had publicly anticipated earlier would be the case. In turn, USDA's August 1995 estimated outlays for farm commodity programs for the fiscal year ending October 1, 1997 (FY1997), were $6.9 billion instead of the $14.4 billion estimated in February 1995, and the estimate for FY1998 was $6.1 billion instead of $13.6 billion.[4] If these new levels of anticipated farm commodity program costs should become the baseline for budget outlays with continuation of current policies, and if the budget cuts worked out between the Agriculture Committees and the Budget Committees were sustained, the new commodity legislation would need to be even more constrained than earlier anticipated in order to meet the lower outlay levels.

—Congressmen Roberts and Bill Barrett, R-Nebr., introduced H.R. 2195, the Freedom to Farm legislation, on August 4. It focused exclusively on commodity legislation and included only four titles—Commodity Credit Corporation Farm Expenditures, Activities in Support of Farming Under Agricultural Act of 1949, Commission of 21st Century Production Agriculture, and Suspension of Certain Provisions Regarding Program Crops. Sen. Thad Cochran's, R-Miss., office signaled that the senator was not accepting Congressman Roberts's proposed Freedom to Farm legislation and stated

that he was opposed to decoupling of payments from commodity prices.[5]
—Congressmen Roberts and Barrett did not include CRP provisions in their proposed Freedom to Farm act legislation. However, they indicated that they desired to maintain the CRP program at 32 million acres. Congressman Roberts was also critical of CBO's continued position that the CBO baseline allowed for only 15 million CRP acres after the year 2000.[6] These CBO numbers became critical points of controversy. The higher the level of acreage CBO assumed would be enrolled in the CRP in any one year, the larger would be the "starting base" of money the Agriculture Committees had for allocating among fulfilling mandated cuts in outlays, continuation of the CRP, commodity programs such as deficiency payments to producers and farmland owners, or other programs over which they had jurisdiction.
—Congressman Roberts also introduced H.R. 2147. It provided that producers would no longer be required to buy catastrophic crop insurance as a condition of participating in commodity programs and receiving program payments. They would have to sign a statement, however, that would waive any rights to disaster aid. The insurance industry opposed this proposed change. In addition Secretary of Agriculture Glickman indicated that he also favored continuation of the then present approach.[7]
—A House Agriculture Committee paper stated that the Agriculture Committees would draft a seven-year farm bill in two phases. One phase would focus on commodity legislation that was to generate savings of $13.4 billion as measured against the February 1995 CBO baseline. The second phase would focus on other programs including credit, trade, and food stamps. According to the Agriculture Committee paper, the farm bill would reauthorize the food stamp program. Thus, the food stamp program would be part of the budget reconciliation bill, and the budget savings associated with its reauthorization would be credited to the Agriculture Committee.[8]
—In the Senate, Richard G. Lugar, R-Ind., and Patrick J. Leahy, D-Vt., introduced a bill, S. 854, to limit enrollment in the CRP to 36.4 million acres in any of the years 1996–2005. They proposed that the CRP emphasize environmental goals. Emphasis on environmental goals was expected, if implemented, to shift CRP acreage sign-up from the Great Plains toward the eastern part of the country. At its inception, the CRP had been used to get land out of production, as well as for its conservation value.
—Senator Cochran introduced the Agricultural Competitiveness Act,

58   CHAPTER FIVE

   S. 1155, with Sen. David Pryor, D-Ark., as its principal cosponsor.⁹ The bill, if enacted, would (1) continue the Acreage Reduction Program, the 0/92 provision, target prices, loans, and payment limitations of current programs, (2) increase the flex acre (nonpaid) acreage from 15 percent to 25 percent, and (3) increase the soybean loan rate to 85 percent of the previous five-year average of prices (approximately five dollars) so long as the increase was budget neutral.¹⁰ The Cochran bill would keep sugar as a no-net-cost program to the federal government, and it would modify the peanut program so that it became a no-net-cost program.¹¹
—Sen. Thomas A. Daschle, D-S.Dak., and colleagues proposed a "targeted marketing loan" approach to farm commodity legislation. It called for replacing the present system of target prices, base acreages, deficiency payments, payment acreages, and other program features with a system of targeted two-tier marketing loans. An initial quantity of commodity would be eligible for a marketing loan at a level above the average market price. Additional production would be eligible for a marketing loan based on average commodity prices. Senator Daschle suggested that $4.00 per bushel for wheat and $2.75 to $3.00 for corn would be the marketing loan level for the initial quantities. A five-year average price would be the level for the additional production.¹² Any loan deficiency payments (or marketing loan gains) would be limited to fifty thousand dollars. In addition, the multiple-entity rule whereby individuals can significantly avoid current limits would be eliminated.¹³ The emphasis was on assistance to middle-size farms.
—When Congress returned from their August recess the Agriculture Committees had explicit instructions from their Budget Committees to save $13.4 billion over the seven years FY1996–FY2002. More specifically they were instructed to revise the mandatory programs (including commodity programs) for which they had jurisdiction so that the resulting estimated outlays over the seven years FY1996–FY2002 would be $13.4 billion less than the early 1995 CBO estimate of $77.5 billion for these programs with unchanged policies.¹⁴

THREE MAJOR PROPOSALS

At the beginning of September there were three major proposals in play. They all called for more flexibility for farm operators in deciding how to use their land resources.

The proposal of Senator Cochran, S. 1155, was the one most closely aligned with the 1990 farm legislation and therefore the programs being implemented in 1995. The key provision of Senator Cochran's proposal that would save outlays was the decrease in the percent of base acreage—from 85 percent to 75 percent—on which producers and farmland owners would receive deficiency payments. These payments would continue to vary inversely with market prices and continue to socialize the downside risk of commodity prices while privatizing the upside movements. Loan deficiency payments—strongly supported by the cotton and rice commodity interests (Senator Cochran's constituent groups)—would continue. These payments, geared to the extent to which market prices dropped below the loan rate, would be available for all production, not just the production on the acreage for which deficiency payments would be paid.

Another approach, Senator Daschle's plan, S. 1256, would purportedly support smaller producers by providing a higher level of deficiency payment for an initial amount of production and a lower level of deficiency payment for production greater than the initial amount. The lack of details associated with the issuance of the press release announcing the proposal suggests that it had been quickly organized, and even the sponsors had not yet had time to focus closely on its content or how it would work.

The most radical of the three proposals was Congressman Roberts's Freedom to Farm proposal, H.R. 2195. It resurrected the early 1980s decoupling concept of former Senator Rudy Boschwitz, R-Minn., a proposal that would have changed the basis for determining the size of checks that producers and farmland owners would receive from the federal government. These checks—decoupled payments—would not be linked to current commodity prices or production. Essentially, Congressman Roberts's proposed legislation said to producers and farmland owners, "The feds will send you checks over the next seven years because you received checks in the past five years." Thus, farmland owners would know the payments that would be associated with their land for the next seven years.

Congressman Roberts thus creatively used the mandate placed on him by the House Republican leadership. He chose to fix total commodity program payments. They would be both the cap and the floor and be the same regardless of market conditions. This approach met with the approval of the House Republican leadership who were fixated on completing a balanced budget package. The fact that the savings to be generated by Congressman Roberts's proposal were being measured against an out-of-date and thus inflated CBO baseline estimate of the cost of continuing the 1990 Farm Act into the future was of no consequence to them. However, Congressman Roberts clearly understood the implications of the bullish commodity markets for any updating of the baseline. So the challenge was to "capture the baseline"—get approval for the policy change before the leadership said that the changes should be measured against a new baseline that reflected updated market expectations and therefore lower farm program outlays with a continuation of the 1990 Farm Act.

With the Roberts approach, the income stability policy that had been considered a virtue by farm groups for years would be tossed on the discard heap. In its place would be a policy and related programs that would assure variability in income from farming activities, albeit at levels perhaps higher than in the past.

COUNTING VOTES IN THE FALL OF 1995   6

When Congress returned from the August 1995 recess there was a strong impetus to move ahead to formulate farm commodity legislation. The summer had provided opportunities to test constituent responses to the various ways that Sen. Richard G. Lugar, R-Ind., Congressman Pat Roberts, R-Kans., and others were proposing to change farm programs, that is, how federal money would be transferred to farm interests in the future. It had also been an opportunity to discern reactions of other legislators and the administration to the alternatives that had been promoted.

Now it was time to begin counting votes among House and Senate Agriculture Committee members in committee markup sessions. Then in each house the committee farm legislation would be combined with legislation approved by other committees. These budget reconciliation acts would be considered on the respective floors, amended, and approved. There would be a conference committee of senators and House members to resolve differences, votes again on the floors, and finally transmission to President Clinton for signature.

The outcome of these coming struggles about farm legislation was unpredictable in the fall of 1995 and, as it turned out, until very near the end in the spring of 1996. Many key events, economic circumstances, political settings, and relationships among participants, which in retrospect were important, were simply not predictable in the fall of 1995.

OVERRIDING CONSIDERATIONS

There were two overriding considerations for most members of the Agriculture Committees that seemed to conflict. First was the challenge to respond to constituents' desires for a continuation (or expansion) of transfers associated with the 1990 Farm Act. The programs undergirding these transfers had evolved over time and reflected the interaction among economic and social conditions affecting farmland owners and operators, their articulated needs and desires, and opportunities for elected officials to respond to these constituents.

The second consideration was that the Republican congressional leadership was determined that the resulting budget reconciliation bill would be a blueprint for achieving a balance in the federal budget in FY2002 and that agriculture commodity programs would be part of the plan. Of utmost importance, however, was the reality that historically the farm program transfers had been closely linked to land and therefore the economic value of the transfers had been capitalized into the value of the land. This reality was, for example, a concern of Senator Lugar's in late 1994 and early 1995 when he first advocated a phaseback of target prices by 3 percent per year that, in turn, would cut deficiency payments paid to landowners and operators of farms and ranches producing the program crops. No one wanted to do anything that might bring about declines in land prices.

There were also two deadlines of importance. The most immediate one for the aggies was the deadline given to the authorizing committees by the Budget Committees. The committees were instructed to decide by September 22 on how farm legislation would be adjusted in order to achieve $13.8 billion of savings in farm commodity program outlays.[1] The benchmark for measuring the savings was to be the February 1995 CBO baseline estimates of how much the FY1996–FY2002 federal outlays would be if the 1990 Farm Act and associated commodity policies and programs were extended into the future. No one was quite sure what might happen if either or both of the Agriculture Committees failed to respond to the instructions of the Budget Committees. However, all who cared knew that the Budget Committees could impose their will if the authorizing committees failed to act.

The other deadline was more remote but nonetheless significant. Most all were of the impression that "current" commodity legislation was to expire at the end of calendar 1995. If the 1990 Farm Act were not extended or substitute legislation were not developed by Congress and signed by the president before January 1, permanent farm commodity legislation from the 1938 and 1949 Farm Acts would then become the law of the land on January 1, 1996. Presumably, such a situation would force the announcement and the implementation of substantially higher price supports and other program provisions that would inflate federal outlays and be grossly incompatible with farm commodity prices and export opportunities.

Thus, the time for legislators to choose among alternative approaches to farm commodity policies was at hand. It was time to count votes.

Voting in the House

Three legislative proposals were the focus of debate among members of the House Agriculture Committee when it met on Wednesday, September 20, 1995, to mark up farm commodity legislation to be included in the House budget reconciliation bill. The legislation included a proposal by the Democrats; a proposal advocated by Congressmen Bill Emerson, R-Mo., and Larry Combest, R-Tex., H.R. 2330; and H.R. 2195, the approach embraced by House Agriculture Committee Chairman Roberts.

The Democrats' Proposal

The proposal put forth by the Democrats included reducing the percentage of acres for which producers would receive deficiency payments from 85 percent to 80 percent. Payment limits would be reduced only slightly. Dairy price support activities would be limited to cheese. Changes in the peanut, sugar, and tobacco programs would be very limited.[2] Outlays over the seven years FY1996–FY2002 with the Democrats' proposal were estimated to be less than if current programs were continued into the future, but not close to the amount necessary to be consistent with the instructions of the House Budget Committee to the Agriculture Committee.

Emerson-Combest

The Emerson-Combest proposal, H.R. 2330, was developed as an alternative to the approach promoted by Congressman Roberts. It was similar to a bill introduced on August 10 in the Senate by Sen. Thad Cochran, R-Miss., S. 1155. H.R. 2330 called for deficiency payments on 70 percent of a producer's base acres. However, it would allow producers to plant all of the base acreage to alternative crops (undefined). According to the CBO estimate, it would generate a $13.4 billion outlay savings relative to the early 1995 CBO baseline. These savings would be related to a combination of decreasing the acreage upon which operators and farmland owners would receive deficiency payments, eliminating the emergency livestock feed program, eliminating the mandatory crop insurance program, changing the peanut program slightly, modifying the 0/85 and 50/85 provisions, reducing funding of the EEP by 20 percent, and limiting the acreage in the CRP.

Freedom to Farm

Congressmen Roberts and Bill Barrett's, R-Nebr., bill, H.R. 2195, called for a radical change in the system used to transfer money to producers of farm commodities and farmland owners. The biggest and the most radical adjustment would be to guarantee payments to producers and farmland owners for seven years based on what they had received in the past. The total payments to all recipients in any one year would be specified, and these payments would be made regardless of the level of commodity prices. Those who owned or operated farmland that had been associated with commodity program payments from the CCC in three of the past five years would be eligible to sign "market transition contracts" to receive payments for the seven crop years 1996 through 2002. The government would agree to make the payments without regard to the level of commodity prices.

The one quite modest requirement for qualifying for the payments would be for recipients to continue previously required conservation practices (conservation compliance) on the farmland with which the payments were associated.

The approach incorporated in H.R. 2195 was obviously in sharp contrast to the prevailing Republican attitude toward the welfare programs implemented by the Department of Health and Human Services. For those programs, attention was concentrated on ways to limit the continuation of transfers to the recipients. Under H.R. 2195 the criteria for receiving payments over the next seven years was whether people had received farm commodity payments in years previous to 1996.[3] If they had, they were eligible to receive payments for each of seven years. If they had not, they were excluded!

Past government programs designed to adjust acreage of crops from year to year would be discontinued. There would be no more efforts to restrain supply by land set-aside (ARPs) determined annually by the secretary of agriculture. However, nonrecourse commodity loans would continue to be available for the program crops (wheat, feed grains, cotton, rice, and oilseeds) as would marketing loan deficiency payments. The loan levels would be set at 70 percent of the olympic average of prices of the past five years.[4] All production of these program crops would be eligible for loans and marketing loan deficiency payments. The nonrecourse commodity loans would be linked with current production, although the guaranteed seven-year contract payments would be decoupled from current market prices and current production. Thus, a safety net at low prices was kept in place. The total guaranteed payments for the seven years would be $38.4 billion. In the first year payments would total $6 billion. In FY2002 these payments would total $4.375 billion, thus declining by about one-fourth over the seven-year period of the proposed legislation. With limited exceptions, producers could grow whatever combination of crops they wished without endangering their payments.

House Committee on Agriculture, Subcommittee on Livestock, Dairy, and Poultry Chairman Steve Gunderson, R-Wis., had been working aggressively during the summer to find a consensus on ways to change the federal dairy programs. However, the regional conflicts among dairy producers made this challenge extremely difficult. On Friday, September 15, 1995, he concluded that it was impossible and proposed that the price support programs be eliminated on January 1, 1996, and that the thirty-eight milk marketing orders be eliminated on June 30, 1996. In addition, he proposed that transition payments be made for five years to "active" farmers. In 1996, milk producers would receive two payments. Payments in the out years would be five to fifteen cents per hundredweight. The payments would be geared in some way to production in three of the past five years. No payments would be made for increases in production or to new entrants into dairying. The assessment on production paid by producers, which in 1995 was fourteen cents per hundredweight, would be eliminated. The proposal referred to as Gunderson II was, in turn, included by the Republicans in Congressmen Roberts and Barrett's Freedom to Farm proposal, H.R. 2195, to be considered by the House Agriculture Committee on Wednesday, September 20, 1995.

This Gunderson II legislation was a substitute for what Congressman Gunderson had been proposing but had failed to attract widespread support—consolidate the then current thirty-eight milk marketing orders into one order with a one-dollar differential between milk used as fluid milk and milk used in manufacturing. Milk prices had been supported through the years by government purchases of manufactured products. The marketing orders facilitated price discrimination in charging consumers a higher price for fluid milk than is paid for milk used in the manufacture of butter, cheese, and dried milk. This separation of the two markets raised dairy producers' incomes.

The House Markup—a Failure

The September 20, 1995, meeting of the Agriculture Committee of the U.S. House of Representatives called by Chairman Roberts to mark up a 1995 farm bill was a failure in many respects. Nonetheless, the meeting of the committee demonstrated (1) the attachment Congressman Roberts had to his Freedom to Farm proposal, (2) cotton and rice interests' steadfast reluctance to embrace Freedom to Farm, and (3) the inability of the Democrats to formulate and accept an approach to 1995 farm legislation that combined required budget restraints with the program approaches to which they had become so attached including target prices, support of selected commodity prices with nonrecourse loans, deficiency payments, and marketing loan payments.

The meeting of the committee was long and did not adjourn until slightly after 10 P.M. It was filled with acerbic debate, infrequent friendly banter, lamentations that farm policy had become partisan, and a series of votes that revealed that a majority of the members of the committee would not support any of the proposals put before it, including the freedom to farm proposal, H.R. 2195, promoted by Chairman Roberts.[5]

The first major vote of the committee was on the Democrats' proposal. The count was 22 yeas and 25 nays. It was scored by CBO as saving $4.4 billion over the seven years 1996–2002 relative to CBO's February 1995 estimate of the cost of the current programs if extended into the future.[6]

The second major vote of the committee was on the Emerson-Combest bill, H.R. 2330. It had a CBO score of $13.4 billion of savings over seven years. It would continue the basic features of current programs. There would be no cap on total payments to producers and farmland owners. In years in which commodity prices were high, the outlays would be less than when prices were low. At the end of the seven years the actual outlays for the seven years could be more or less than estimated by CBO in 1995. This proposal was also narrowly defeated, the count 23 yeas and 26 nays.

The third major vote was on Congressman Roberts's Freedom to Farm proposal, H.R. 2195. The time was about 9:30 in the evening. It also had a CBO score of between $13 and $14 billion. The vote was 22 yeas and 25 nays. Congressman Gunderson initially voted yea, but then changed his vote from yea to nay so that he could move to reconsider if that procedure became useful. Four Republicans on the committee including Congressmen Bill Emerson, R-Mo., Larry Combest, R-Tex., Saxby Chambliss, R-Ga., and Richard H. Baker, R-La., voted nay.

A shocked chairman with the words, "The committee will be in recess until 10 o'clock! That is 10 o'clock tonight!" recessed the committee while he sought to cajole the four Republicans who had voted against him to change their votes to yeas. But his efforts were of no avail. They would not change, yet. The chairman returned to the podium at the center of the dais in the Agriculture Committee hearing room in the Longworth Office Building, banged the gavel, and declared the House Agriculture Committee adjourned until the call of the chairman. The many, many lobbyists for those with a financial stake in the farm bill, the limited number of members of the press corps, the few observers, the USDA representatives, the members of the staffs for the committee and committee members, and members themselves—all equally flabbergasted at the outcome and the unusual abrupt ending—quietly gasped and looked at each other. They slowly walked away wondering if what they had observed had actually happened and whether the House leadership's threats about farm legislation would in fact happen—namely that the House leadership would write Congressman Roberts's freedom to farm proposal into the budget reconciliation bill even if the House Agriculture Committee had not approved it by a simple majority.

Through the day there had been other votes on amendments to the freedom to farm proposal. Only three passed. One of these called for $200 million each year for water and waste grants. The other two were technical amendments to the peanut and sugar provisions.[7]

Congressman Roberts reconvened the committee the following week to announce that he would no longer pursue efforts for the committee to report out a farm bill that could be included in the prospective House budget reconciliation bill.[8] Thus, the House leadership seemed to have decided to deliver on the threat that was included in two September 14, 1995, letters from the House leadership to Congressman Roberts. These letters were widely interpreted to mean that if the House Agriculture Committee did not approve Congressman Roberts's proposed legislation, the leadership would merely insert Roberts's proposed farm legislation into the prospective omnibus budget reconciliation bill.

One letter was signed by Congressman John R. Kasich, R-Ohio, as chairman of the House Budget Committee. It stated that if authorizing committees do not report changes in law "such that outlays for programs under their jurisdiction do not exceed a specified level...the House Committee on the Budget will offer legislative language to the House Committee on Rules to bring the bill into compliance with the reconciliation instructions."[9]

The second letter was signed by Congressmen Richard K. Armey, R-Tex., Newt Gingrich, R-Ga., and Tom DeLay, R-Tex. It expressed hope that the Agriculture Committee would approve freedom to farm and stated that, "It is consistent with the goals of the new Republican Congress." In addition, it stated that they would consider two alternatives "if the committee fails to report such reforms..." One would be to bring "a farm bill to the floor ... under an open rule." The other would be to "replace the committee's legislation with true reforms before reconciliation is considered on the floor."[10]

The primary reason why the House leadership endorsed Congressmen Roberts and Barrett's proposed Freedom to Farm bill seems to have been associated with a characteristic of the proposal—*change*—instead of any great analysis of its advantages or disadvantages, for there was no such analysis. The proposal did constitute change. Further, the House leadership had concluded, rightly or wrongly, that this particular change constituted reform. The fact that its cost was fixed was promoted as a virtue. However, such a claim of virtue ignores, from a budget viewpoint, the merits of an alternative, namely, to gear the payments to market conditions but place a maximum on total payments that could be made in each of the years. Such an approach would save federal outlays in years of high commodity prices. It would also tend to stabilize farm-related income of operators and landowners by providing assistance in years of low commodity prices. The total amount of assistance, however, would not be open-ended as with the previous farm commodity programs.

Tension among House Republicans

There was much consternation among House legislators subsequent to the failed markup attempt by the House Agriculture Committee. The tension was particularly acute among the Republicans. The Democrats recognized that the Republicans controlled the farm-related activities and that any efforts to oppose the inclusion of Freedom to Farm in the House reconciliation bill would be futile.

Several developments illustrate the tension among Republicans, including

—A memo was circulated that described possible ways that the four Republicans who voted against Roberts in the committee markup could be disciplined.[11] At the same time some individuals in the leadership ranks hinted at making adjustments to Freedom to Farm in response to the cotton and rice interests represented by the four resisters.[12] Others in the House leadership affirmed the plans to insert the Freedom to Farm bill into the House reconciliation bill. In the meantime, the four Republican Freedom to Farm resisters (Congressmen Emerson, Combest, Chambliss, and Baker) sent a letter to their House colleagues. In it they threatened to vote against the reconciliation bill and criticized Congressmen Roberts and Barrett's Freedom to Farm proposal as welfare.[13]

—Congressman Gerald B. H. Solomon, R-N.Y., chairman of the House Rules Committee, objected to the dairy provisions in Freedom to Farm as being unduly favorable to the Midwest.[14] He and other northeastern members of Congress coordinated activities with several southern members of Congress in opposing the freedom to milk provisions included at that time in Congressman Roberts's Freedom to Farm proposal.

—Ten southern Republican House members (other than the four resisters) and a Democrat asked Speaker Gingrich to place the Senate Agriculture Committee-approved farm bill in the House reconciliation bill instead of Freedom to Farm. In their letter they emphasized that federal outlays with Freedom to Farm would be larger than with the then Senate bill and indicated their preference for farm programs that "assist farmers in years of lower prices."[15]

In the end, the Republican objectors to Freedom to Farm were placated with an offer that their concerns would be addressed in the prospective House-Senate conference on the reconciliation bill. In reality, reports of a meeting of Speaker Gingrich with the House Republicans suggest that Speaker Gingrich's call for party discipline in moving the balanced budget legislation was the primary factor that muffled House Republican objections to Freedom to Farm and maintained the Republican votes in support of the reconciliation bill.[16] There were probably several reasons why the leadership continued to embrace Freedom to Farm. Three likely ones were (1) to change support from Freedom to Farm to some other approach would undercut Congressman Roberts, (2) to deviate from Freedom to Farm as brought to them by Congressman Roberts could enmesh the leadership in the intricacies of farm programs at a time when they were already overburdened with trying to complete the budget reconciliation bill, and, possibly even more importantly, (3) to adjust the farm portion of the bill could add to pressures to adjust other sections of the budget reconciliation bill.

Thus, the House budget reconciliation bill containing Freedom to Farm went to the floor in late October. The House vote occurred on October 26, 1995. The bill passed with 227 yeas and 203 nays. All of the dairy dissenters and the rice and cotton dissenters voted yea. Party discipline prevailed. Thus, Congressmen Roberts and Barrett's Freedom to Farm proposal was approved by the House as part of the much larger reconciliation bill and went for conference between the House and the Senate, even though Freedom to Farm had not been able to attract a majority vote in the House Agriculture Committee. Of course, no other proposal had received a majority vote either.

Voting in the Senate

In the meantime the Senate was developing its budget reconciliation bill, which included farm commodity legislation.

The Senate Markup Votes

Two proposals were the focus of the votes among members of the Senate Agriculture Committee when it met on September 27—a proposal by the Democrats and a proposal around which Senator Lugar eventually found sufficient agreement among the Republican members to report it out of the Agriculture Committee.[17]

Under the Democrats' bill, price protection for each participant in the program would be offered for limited preset amounts of a commodity including, for example, up to 22,000 bushels of corn at $2.75 and up to 15,400 bushels of wheat at $4.00. Wheat, feed grains, and oilseed marketing loans would, under the Democrats' plan, be 95 percent of the previous five-year average. Cotton and rice nonpay acres would increase from 15 percent to 22.5 percent of base acreages. Other provisions would make limited adjustments in the sugar, wool, and dairy programs.

The limit on the quantity of a program commodity eligible for payment could serve to limit payments to larger producers and landowners, but that would depend on how the program implementations rules were constructed. Rules which permitted proliferation of partnerships would perpetuate the circumscribing of payment limitations of past legislation.

The Democrats' proposal was scored at $4.2 billion of savings for commodity and conservation programs.

Senator Lugar's approach now included the proposal fostered by Senator Cochran. It involved increasing the normal flex acres to 30 percent of base acreages from its current 15 percent. To receive benefits the recipients would have had to have been enrolled with at least one crop base in three of the past five years. The emergency feed program would be eliminated, dairy supports would be confined to cheese, funds for EEP and the Market Promotion Program (MPP) would be reduced, and a new Environmental Quality Incentive Program (EQIP) would be initiated. The CBO budget scores were $13.4 billion for the conservation and commodity programs and $35.7 for nutrition programs.

It took two days for the Senate Agriculture Committee to vote out a bill for inclusion in the Senate's budget reconciliation act. On the first day the Republicans were stalemated on reporting out their bill. Rick Santorum, R-Pa., voted with the Democrats and against the bill favored by the chairman, Senator Lugar. The vote was a tie, 9 yeas and 9 nays.

Senator Santorum was opposed to the sugar and peanut provisions included in Senator Lugar's bill. However, the Democrats' plan was also defeated. This time the vote was 8 yeas and 10 nays. On this vote Senator Santorum voted with his Republican colleagues.

By the second day, September 28, Senator Santorum had obtained some concessions from the Republicans on peanuts and sugar. These changes were sufficient for him to vote "present." Thus, by a vote of 9 yeas and 8 nays the Senate Republicans' commodity plan was destined to be become part of the Senate's budget reconciliation bill.

Senate Approval

On Friday, October 27, 1995, the Senate considered their budget reconciliation bill. Sen. Max Baucus, D-Mont., offered an amendment related to farm commodities. The amendment had a CBO score of $4.4 billion of savings over seven years in comparison to the $13.4 billion saving score associated with the Agriculture Committee proposal included in the Senate's budget reconciliation bill. The Baucus amendment called for fewer federal tax cuts as an offset to the commodity programs being cut by $4.4 billion instead of $13.4 billion. The amendment failed with the vote of 46 yeas and 53 nays.[18] The Senate stayed in session until the early minutes of Saturday morning, October 28, 1995, when it approved a budget reconciliation bill containing the Agriculture Committee-approved farm commodity title by a vote of 52 yeas and 47 nays.

FREEDOM TO FARM AS PART OF BUDGET RECONCILIATION

House-Senate Conferencing

As the fall of 1995 slipped by, farm legislation activities continued to be held captive to the exhausting tasks associated with finalizing mammoth-size budget reconciliation legislation to submit to the president for what was an increasingly likely veto. Designated conferees to resolve farm bill differences between the Senate and House included both Republicans and Democrats. However, it was widely known that the decisions were largely, if not exclusively, made by the Republicans on the House side, in some cases with the exclusion of Republicans who had not aligned their support with the Roberts proposal.

There were fundamental differences to be resolved. Senator Cochran's approach was to achieve calculated budget savings within the then current program structures. The House approach would continue to make transfers to producers and farmland owners in the years ahead, but base the transfers largely on the transfers that had been made to these individuals and entities over the past five years.

Both bills called for reductions in commodity program outlays. However, there were additional differences. These differences related particularly to how payments were determined and to commodity price supports. The Senate bill would assure that there would be outlay savings relative to continuing the 1990 Farm Act into the future. These savings would be realized by decreasing the number of acres that would qualify for deficiency payments.

Also with the Senate bill, farm payments would assuredly be less than payments would be with a continuation of current policy. But their exact level in any one year would depend on commodity market prices in that future year. Therefore, outlays in coming years could exceed the fall 1995 expected outlays. This countercyclical characteristic had been a vital feature of farm programs for decades. Transfers would be more if prices declined and less if prices were higher. The reasoning was that transfers from the government ought to be higher when market prices are depressed and less when market prices are high.

In contrast to the Senate bill that went to conference, the House proposal could lead to outlays greater than an extension of the 1990 Farm Act. If market prices were high, outlays under 1990 legislation would be less or even zero. However, under the House bill the outlays would be the fixed amount set in the legislation. Alternatively, if low prices developed, outlays under the House bill would not exceed that allowed in the new legislation unless, of course, the law was changed.

In the fall of 1995 it looked increasingly likely that federal outlays in the coming seven years with a continuation of USDA farm commodity-related programs would be lower than they had averaged over the previous few years. However, comparisons of prospective budget outlays over the seven years (1996–2002) with a continuation of existing policies with prospective outlays over the same years with the different proposed policies caused much confusion during this period. One reason for the confusion was that the budget procedures called for using early 1995 CBO forecasts of 1996–2002 market conditions throughout the 1995 legislative debate. These particular forecasts were, of course, based on fall 1994 commodity market prospects. On that basis the outlay savings associated with both the Senate and the House proposals for inclusion in the congressional budget reconciliation bill would be $13 to $14 billion over the seven years (1996–2002) compared to the estimated outlays if the 1990 Farm Act were continued for the years FY1996–FY2002.

The savings arithmetic drastically changed from late 1994 to late

1995. For example, Bruce Gardner, a former Assistant Secretary of Agriculture, concluded on the basis of improved market conditions (October 1995) that if the 1990 Farm Act were simply extended, 1996–2002 outlays would be $40 billion and not $56 billion as was estimated by CBO in late 1994–early 1995. In comparison, the particular legislation approved by the House of Representatives would lock in $43 billion of outlays. Gardner concluded, "Therefore, the FFA [Freedom to Farm Act] gets credit for $13 billion in farm program outlay reductions while actually spending several billion more than would be spent under current law!"[19]

Emerging Prospects

As the Senate and House conferees continued to pursue an agreed set of farm commodity provisions for the prospective congressional budget reconciliation bill, it became increasingly likely that with the resulting agreed legislation

- —operators and farmland owners would have greater flexibility in making planting decisions,
- —producers would be expected to meet conservation compliance provisions,
- —participation in crop insurance programs would not be a condition for receiving commodity program benefits,
- —acreage devoted to the principal U.S. crops over the next seven years would be larger than in recent years,
- —funding of the EEP would be less than it had been in recent years, and
- —producers and farmland owners would bear greater price risks. However, if the House approach were accepted, there would be contracts assuring operators and farmland owners of specific transfers in each of the next seven years. These transfers would go a long way in offsetting the increased price risks, perhaps more than offsetting them.

Several different notions on how to devise a compromise in conference were proposed. One approach combined long-term contracts, nonrefundable deficiency payments, 30 percent unpaid acreage, and planting flexibility. Another involved combining Freedom to Farm-type fixed payments with a payment based on market conditions. Still another would have given producers and landowners a choice between Freedom to Farm or an approach akin to current legislation.[20]

Conference Agreement

On November 9, 1995, Congressman Roberts announced that the conferees had agreed to go forward with Freedom to Farm.[21] It was revealed later that major concessions had been made to the cotton and rice interests, two groups that had opposed Freedom to Farm. Loan rates were to be 85 percent of the past five-year average, not 70 percent as had been included in Congressmen Roberts and Barrett's August 1995 Freedom to Farm proposal.

A November 15, 1995, news release for both the Senate and House Agriculture Committees detailed the agreements reached by the conferees. They included

—elimination of the ARP,
—elimination of permanent law provisions,
—elimination of the mandatory purchase of crop insurance requirement,
—elimination of the farmer-owned reserve, and
—elimination of target prices.

The law would set maximums for loan rates at the 1995 levels. Minimums for cotton would be $.50 and for rice $6.50. Within these constraints the loan rates would be set at 85 percent of a five-year average.[22]

There were other compromises. The three-entity rule would continue (the House bill would have eliminated it), marketing loans would continue, the cotton step two payments would continue with the current 1.25 cent/pound trigger. However, the eight-month cotton loan extension would be eliminated and there would be a 1 percent increase in the CCC loan interest rate.

The total outlays for contract payments in the conference bill were reduced from $38.7 billion (the amount that had been included in the House-passed reconciliation bill) to $35.6 billion.[23]

Congressman Roberts's political strategy was somewhat revealed in a November 21 press release. The release included specific dollar figures for "prospective market transition payments" that would be made to operators and farmland owners if the conference bill were approved and for USDA's estimated deficiency payments for crop year 1996 if the then current law (1990 Farm Act) were extended. Given the high commodity price prospects, these USDA-estimated deficiency payments were lower than the "market transition" payments under the proposed Freedom to Farm program and demonstrated that Roberts's program was more generous to producers.[24]

Further, Congressman Roberts took opportunities to mention that with then current law, operators and farmland owners were required to repay the advanced deficiency payments they had received for the 1995 crop. Commodity prices had increased to the point that producers were not entitled to deficiency payments. Thus, producers were now expected to repay the advanced deficiency payments they had received when prices were lower. It was obvious that Congressman Roberts was trying to find a way to ease this obligation of operators and farmland owners. Because market prices were up, a continuation of existing policy would not provide that relief.

With the conclusion of the conference of the House and Senate aggies, the stage was set for the budget reconciliation bill to include a Freedom to Farm commodity program akin to what Congressmen Roberts and Barrett had proposed in August 1995. However, there were a limited number of major differences between Congressman Roberts's earlier proposal and the measure agreed to in conference. The major difference was the setting of support prices at higher levels. Other differences included the continuation of complex provisions related to cotton that had been devised over time to increase the transfers from the federal government to cotton interests.

The budget reconciliation legislation, including the farm commodity title, passed both the House and the Senate on November 28, 1995. Thus, the next immediate question for the aggies was whether the budget reconciliation bill approved by the Congress would be signed or vetoed by the president. It was not a long wait to find out. On December 6, 1995, he vetoed it.[25] The president mentioned the farm part of the bill in his veto message. However, other parts of the budget reconciliation bill clearly accounted for his veto. No one knew whether he would have signed or vetoed the farm legislation if it had gone to him as separate legislation.

A Summary of What Happened

There was a sharp contrast between what happened in the House and in the Senate Agriculture Committees as preparation of their respective budget reconciliation bills continued in the fall of 1995. In the House Agriculture Committee none of the proposals put forth attracted a sufficient number of votes for the committee to report out a bill for inclusion in the reconciliation bill, not even the Freedom to Farm approach embraced and promoted by the chairman, Congressman Roberts. In turn, the House leadership did as Congressman Roberts had arranged.

It included Congressmen Roberts and Barrett's Freedom to Farm bill in the House budget reconciliation bill that went to the floor of the House for a vote. It was approved under a closed rule, that is, no debate.

In the Senate, Senator Lugar backed off from his proposal to cut target prices by 3 percent per year over a series of years. Instead, he accepted Senator Cochran's proposal, which generated budget savings by decreasing the number of crop acres on which deficiency payments were to be paid. It was reported out by the committee and included in the Senate's budget reconciliation bill. In turn, it was approved on the floor.

In the House-Senate conference on the farm part of the reconciliation bill, however, Senator Lugar transferred his support to the House-proposed Freedom to Farm. With important changes, notably the setting of support prices at 85 percent instead of 70 percent of the average price of the past five years, Freedom to Farm was approved by the conference and became part of the budget reconciliation bill that was forwarded to the president and vetoed on December 6, 1995. However, by that time Freedom to Farm had sufficient support in Congress and among agricultural constituents for it to become in March 1996 a component of a classic logroll directed by Sen. Patrick J. Leahy, D-Vt., and Senator Lugar. It ended up as the commodity section of the 1996 Farm Act.

Continued strong commodity markets reinforced optimism over future commodity prices. This optimism undoubtedly increased the number of operators and farmland owners (and their proxies) who were attracted to programs that promised payments not tied to prices. With strong prices, programs that promised benefits only when commodity prices were depressed would be of little use in generating transfers from the federal government to operators and farmland owners. The extension of this thinking was that such provisions would not pay out in the future. Consequently, producers would be better off (and taxpayers worse off) if somehow the programs could be constructed so that transfers were made even if commodity prices were high in the future. Among the agricultural interests a horse trade became increasingly attractive—abandon past approaches which tied commodity program payments to commodity prices (the lower the market price, the higher the transfers) in exchange for a system of transfers that would occur regardless of the level of commodity prices. The arcane commodity program provisions, the budget process, and the lag of CBO scoring taking into account the increased market prices bolstered the probabilities of this horse trade.

Thus, the September 1995 presumed conflict between meeting budget cuts and continuing transfers of the magnitude to which producers and farmland owners had become accustomed was seemingly avoided. There were two critical situations that contributed to this development. One was strong commodity prices. The other was CBO lags in updating their outlay estimates associated with a continuation of current programs. In addition, the inclusion of the elimination of the ARP in the Freedom to Farm approach made it possible for the agribusiness interests to attain their goal of increased volumes of inputs and products to handle and process without appearing to be in conflict with the financial interests of producers or farmland owners. The interests of the public with respect to consumer prices and tax implications of farm legislation were essentially ignored except in the overwhelming macrodimensions of balancing the budget, lowering taxes, and having payments to agriculture that were firmly budgeted in advance.

A Wait and See Season   7

President Clinton's December 6, 1995, veto of the budget reconciliation bill changed the atmosphere surrounding prospective commodity legislation. In many respects the veto marked the beginning of a "wait and see" season for the aggies on the Hill—for both the elected and their staffs, as well as for those hired to lobby the elected on behalf of their constituents. This season lasted until near the time the ensuing budget summit faded away, some six to eight weeks later.

Four conditions contributed to the wait and see atmosphere for agriculture

—the nation's focus on the veto and the ensuing budget summit,
—no shutdown of USDA as happened to several other federal agencies,
—low probabilities of permanent commodity legislation being invoked with the expiration of current programs, and
—the inability to extricate commodity legislation from the larger budget negotiations.

Together, these conditions generated considerable nervousness about what the outcome of the budget summit would be for farm commodity policy in 1996.

Attention on Budget Summit

Press attention to the jockeying over whether there would be a budget summit, the preparations for it, and its actual conduct eclipsed the public visibility of activities related to farm commodity legislation. The farm provisions of the reconciliation bill were of secondary interest to Congress and to the administration. Education, welfare, environment, health care, and once in a while "corporate welfare" were the mainstays in the larger debate over how the federal budget might be balanced sometime in the future. These topics, not agriculture, attracted the public's attention, including most of those with interests in agriculture.

No USDA Shutdown

The Republican leadership decided to further delay several appropriations bills for the fiscal year that had begun on October 1, 1995. The related threat of shutting down significant parts of the government was to be leverage on the president to accommodate budget reconciliation legislation. However, the employment of this tactic merely reinforced the ongoing lull regarding farm commodity legislation. USDA was one of the eight appropriations bills that had been approved by Congress and signed by the president previous to the shutdowns. Consequently, the partial shutdown of the federal government did not direct attention by either the Hill or the administration to farm programs. The twelve agencies shut down did not have their FY1996 appropriations. USDA did and therefore it continued to do those things expected of it. In fact, the press attention to the twelve shutdown agencies and the lack of their services crowded out news regarding the continued delay on farm commodity legislation.

Invoking Permanent Commodity Legislation Not Likely

Only limited attention was given to the December 31, 1995, expiration of the 1990 farm legislation which had governed the commodity programs for the 1991 through 1995 crops. If substitute legislation were not approved, the Agricultural Act of 1949, permanent legislation, would presumably become the legislation controlling commodity programs. If in fact this legislation was controlling and its dictates were followed, the price support levels for most supported commodities would be increased substantially. As a consequence the government's involvement in commodity markets would increase sharply. However, the threat of the 1949 act being implemented was widely discounted, although not dismissed completely.

One reason for discounting the possibility of implementation of the 1949 act was the belief by some people that the secretary, because of a provision in the 1990 Farm Act, already had the authority to declare that the 1990 act was extended into the future. For example, Reuters reported on November 30 that come December 31 the secretary could extend current policy; but, if he did not, the 1949 act would become the controlling legislative authority for farm commodity programs.[1]

Another reason for the limited attention to the expiration of the 1990 Farm Act was that commodity price prospects were relatively bullish and therefore there was no pressure for the secretary to withhold land from production in 1996. Strong pressure groups had already lobbied for an end to annual cropland retirement programs. This lobbying and the bullish market conditions made it unlikely that there would be annual withholding of land from production in 1996 even if the secretary had the authority to do so.

If prices had been depressed, the situation would have been starkly different. At least some groups would have been searching for legal authority for the secretary to declare that a portion of U.S. cropland should be withheld from production in 1996. Those agribusiness groups lobbying for the end of annual land withholding programs would have been on the defensive.

Further, as was evident with the 1996 winter wheat crop, it is possible for producers to plant crops without knowing what the policy would be when the crops were harvested. In the then current market conditions it was clear that the sky was not going to fall if the federal government program for the 1996 crop was not clarified soon. If wheat producers could plant crops without knowing what the eventual program would be, surely other producers could do so also. Admittedly, policy advocates sometimes invoke the notion that farmers must know the prospective applicable policy before they plant the crop. But in the last few months of 1995 and even into the early months of 1996, the optimism over future commodity market prices outweighed by large margins any disruptive effects on producers from not knowing what the commodity policies might be for the 1996 crop.

No Way to Extricate Commodity Legislation from Budget Summit

Since the budget reconciliation bill included a farm title, it was politically impossible for that part of the bill to be immediately resurrected in Congress as separate legislation even though the 1990 Farm Act was due to expire on New Year's eve. There was no farm crisis justifying even a partial dismemberment of the budget reconciliation bill at this time in order to rush through a separate farm bill. Consequently, there was no alternative for the aggies but to observe the summit political activities and wait.

Nervousness over the Budget Summit

The nation's attention was on the budget duel between the congressional Republicans and the White House. It was not clear what the outcome of this duel might be nor the implications of the negotiations for farm commodity legislation. The Republican approach to balancing the budget might fail. In that case the Agriculture Committees would of necessity revisit their farm bill activities. They, of course, could vote a simple extension of the 1990 Farm Act or they might choose to somehow construct new farm commodity legislation. That is, unless the administration beat them to the punch with a simple extension of the 1990 legislation based on interpretation of Section 1129 of the 1990 Farm Act.

On the other hand, the possibility of a White House-congressional leadership deal on the budget could not be ruled out even though the probabilities were low. In such an eventuality, farm legislation could be swept into place with the flurry of activity surrounding agreed budget reconciliation legislation. The activities could have resulted in a simple acceptance of the commodity title contained in the congressionally approved budget reconciliation bill. In that case, Congressmen Pat Roberts, R-Kans., and Bill Barrett's, R-Nebr., Freedom to Farm approach would have become law even though neither Agriculture Committee had voted it out of committee. It had been approved by the full Senate and by the House. Nonetheless, it was still only one title in a very large piece of budget reconciliation legislation which had passed the Senate and House with strict limits on debate and amendments.

In contrast, differences could have arisen over the farm title even if both the congressional leadership and the president had concluded that there was a political imperative to conclude a budget reconciliation bill. Then any differences would have had to be resolved quickly in some way. With that turn of events, the particular preferences of one or two individuals could have led to the substitution of a greatly different approach to farm commodity policy such as a two-year extension of the 1990 act. The big uncertainty with this possible scenario was just how the compromise would be structured.

Only a limited number of key people including the president, House Speaker Newt Gingrich, R-Ga., Minority Leader Richard A. Gephardt, D-Mo., Majority Leader Robert Dole, R-Kans., and Minority Leader Thomas A. Daschle, D-S.Dak., would be involved directly in making a bargain. Such bargaining, because of time pressures, could not have been easily influenced by either of the Agriculture Committees, the secretary of agriculture, or the farm or agribusiness groups with significant stakes in farm legislation. Nor could a bargain on farm commodities have been easily modified by these stakeholders after it became part of a White House-congressional budget deal.

Once it became obvious that there was not going to be any agreement between the White House and Congress on budget reconciliation, there was the inevitable separation of commodity legislation from the discipline associated with budget reconciliation procedures. This breakdown in budget discipline allowed other forces to become the major drivers of the process. The Agriculture Committees gradually took back their traditional roles of setting the agenda for the farm bill debate, voting, compromising, and logrolling. Concerns for real outlay savings and activities of institutions primarily embracing that objective, like the Senate and House Budget Committees, receded into the background and became less and less relevant to the farm bill debate.

Farm Bill Preliminaries Nonetheless

8

The major political focus during December 1995 and January 1996 was on the budget summit. Nonetheless, maneuvering continued on farm commodity legislation, although at a subdued level of intensity. In the end, these activities in varying ways helped set the stage for what would eventually come to pass in the House and Senate Agriculture Committees, on the floors of both houses of Congress, and on the president's desk on April 4, 1996. Most observers recognized that the farm commodity legislation might again reemerge on the aggies' agendas. If the budget summit failed, it was important to be ready to wrap up farm commodity legislation.

In December 1995 and early 1996 the administration continued to be restrained in expressing its desires and refrained from making demands on Congress. Interest groups, farm and nonfarm, seized opportunities to articulate their positions and influence decisions.

In the House, Congressman Pat Roberts, R-Kans., and his Republican Agriculture Committee colleagues prepared for a possible farm legislative battle, should it come, with much more success than did the Democrats. In the Senate, Sen. Richard G. Lugar, R-Ind., in concert with Sen. Patrick J. Leahy, D-Vt., gradually emerged as the linchpin between Freedom to Farm, which he (Senator Lugar) had signed onto during budget reconciliation committee conferencing, and agriculture noncommodity legislation including conservation, food and nutrition, credit, and research.

The story most important to the eventual outcome of 1996 commodity legislation was not occurring in Washington, D.C., but in the farm commodity exchange pits in Kansas City, Minneapolis, and Chicago, and in cereal markets around the world. For the bullish grain markets made the Freedom to Farm legislative proposal increasingly financially attractive to producers and farmland owners and a continuation of the 1990 Farm Act or any variation thereof increasingly less attractive. It was increasingly apparent to operators and farmland owners that under an extension of the 1990 Farm Act payments would be less than they would be if Freedom to Farm were the law.

THE ADMINISTRATION

During the December 1995–January 1996 period the administration continued its low-profile approach to farm legislation. The weeks during which the president was engaged in high-stakes budget negotiations with key congressional leaders were not a propitious time for the administration to press aggressively on commodity legislation. Any move that went much beyond what had already been stated in earlier weeks could undercut the president. However, from time to time the secretary of agriculture indicated a preference for this or that legislative proposal.

For example, in late November Secretary of Agriculture Dan Glickman claimed that if a farm bill was not completed by the end of December, an extension of current legislation would be necessary in order to maintain farm policy and to provide information necessary for producers to make decisions regarding the 1996 crop.[1]

The administration was evidently not searching for a way to simply proclaim an extension of the 1990 Farm Act and thereby continue the farm commodity programs that were authorized by that act. There were not any intimations by department officials that they were seriously considering that option. Instead, USDA then publicized a legal opinion by the department's Office of the General Counsel which stated to the effect that the administration did not have the authority to extend the 1990 Farm Act.

However, the department's opinion ran directly contrary to an opinion of the Congressional Research Service legal counsel that was publicized much later. These conflicting opinions plus the lay interpretation of Section 1129 of the 1990 Farm Act and the floor discussion when the 1990 legislation passed as reported in the *Congressional Record* suggest that if the administration had seriously wanted to effect an extension, they could have announced one, referred to the 1990 discussion, and, in so doing, challenged Congress to either accept an extension or develop an equally effective program with lower expected outlays.

Secretary Glickman signaled in early December his recognition that the administration was unable to compete with the attractiveness of Freedom to Farm. He did this by combining criticism of payments that did not change with commodity prices with a suggested compromise. His compromise approach would, like Freedom to Farm, "capture the baseline." The term "capture the baseline" was shorthand for completing farm legislation in a manner whereby its estimated outlays were measured against the early 1995 CBO estimate of outlays with a continuation of the 1990 Farm Act rather than a CBO estimate that took into account the bullish commodity markets later in 1995.

The legislation suggested by the secretary would require that actual outlays in each year be equal to the CBO-estimated (in February 1995) baseline amounts for the years FY1996–FY2002 less the budget savings as specified earlier in 1995 by the Budget Committees. However, rather than these outlays being distributed each year to commodity producers and farmland owners without regard to market prices for the particular commodities for that year, payments to producers and landowners would be conditioned by commodity prices. The higher the prices, the smaller would be the outlays made to producers and landowners. The remaining money would be used for other programs including rural development and agricultural research. Thus, with Secretary Glickman's proposed approach, the CBO baseline would be "captured" for rural interests, but not just for those people in rural America that produce particular commodities that have been traditionally supported. This approach was vehemently opposed by the traditional commodity groups and their supporters. They did not want others sharing in what commodity interests saw as theirs by right.

In another move, also in early December, the administration sketchily advanced information about an alternative approach to balancing the federal budget in seven years. As part of this alternative the administration called for a $5 billion cut in farm program spending over seven years (relative to the administration's baseline numbers). These savings were to be partially achieved by increasing percentage of land on which payments were not being made from 15 percent to 21 percent. In addition, the administration's approach called for prohibiting payments to people who earn one hundred thousand dollars or more in off-farm income, terminating the tobacco program, and making changes in the peanut program so that the program would not require budget outlays. The administration was not specific regarding dairy, except to claim that changes in that policy would generate $800 million of savings in outlays relative to the administration's baseline.[2]

At the same time, the administration was challenged to keep harmony on farm legislation with the Democrats in Congress. In the House, Congressman Charles W. Stenholm, D-Tex., was a strong voice among the "blue dogs" who were pressing the administration to be more forthcoming in taking steps that would balance the budget. And Sen. Thomas A. Daschle, D-S.Dak., was known to be miffed with the administration over its farm policy stance and in particular the omission of marketing loans in the administration's December 1995 budget proposal. Marketing loans were part of the two-tier proposal advanced by Senator Daschle and other upper Midwest Democrats. Under their proposal producers would receive for a part of their production a marketing loan set at current target prices. Their remaining production would be eligible for a lower marketing loan set at 95 percent of the average market prices over the past five years.

In response to Senator Daschle's concerns, Leon Panetta, President Clinton's chief of staff, sent a letter saying that the administration would consider marketing loans in preparing a final budget package. However, Mr. Panetta did not endorse the two-tier approach advanced by Senator Daschle, nor did the administration do so later.[3]

Senator Daschle's December 1995 push for endorsement of marketing loans was not new. In August 1995 Senator Daschle and a limited number of Senate Democrats had proposed a two-tier marketing loan program which would have placed limits on the benefits to larger producers to the advantage of smaller and medium-size producers. The proposal in August, although put forward without much detail, called for marketing loans for a limited quantity at target price levels and marketing loans at 95 percent of five-year average market prices for the remainder as did the concepts advanced by Senator Daschle in December.

The administration continued to deal with the question as to whether to plan for an extension of the 1990 Farm Act for the 1996 crops and dairy production either as a result of a legislative initiative or by invoking the authority of Section 1129 of the 1990 Farm Act. Sen. Byron L. Dorgan, D-N.Dak., and some other congressional Democrats saw advantages in the prospect of the president extending the 1990 Farm Act into the future. The administration, however, obviously veered away from that course. The department wanted changes in the farm legislation and evidently either considered Freedom to Farm preferable to a continuation of current legislation without modification or were optimistic that they could somehow bring about a modification of Freedom to Farm to make it more to their liking.

The administration reportedly discussed the possibility of submitting legislation to Congress that would continue the basic features of the 1990 Farm Act, but with changes to facilitate greater cropping flexibility for producers. They never did this.[4] Instead, the secretary continued to hope to develop compromises with Congress, including, as discussed above, the possibility of changing Freedom to Farm so that the amount of direct payments would be reduced and the money saved by this modification would be used for other USDA programs such as research and rural development.

In Congress

During January 1996 it became increasingly evident that the president was outmaneuvering congressional Republicans on the balanced budget issue. Shutting down the government with delays in approving appropriations as a way to pressure the president on budget issues met with public disapproval. The president mastered arcane budget detail and maneuvered the Republicans into extended meetings. And the Republicans became exasperated as they were continually frustrated by President Clinton's approach to the discussions. In the end they were unable to generate sufficient concern by the electorate over the balanced budget issue to cause the president and his staff to adjust their tactics. The administration leaders had accurately anticipated that their strongest political position was to refuse to sign onto the congressional approach to the budget. Thus, the probability of the White House-congressional budget summit finalizing the 1996 farm bill, while never high, became nil.

Congressman Roberts continued to promote the Freedom to Farm approach. On January 5, 1996, he introduced it in the House as freestanding legislation. The failure of the White House-congressional budget talks offered the prospects that the Agriculture Committees could fashion a 1996 farm bill with less interference by the Budget Committees than was the case during the preparation of the then defunct budget reconciliation act.

There were perils associated with moving ahead to a freestanding bill. Opponents of the bill would likely have an opportunity to amend the legislation on the floor—something that was prevented with the rules for floor action on the budget reconciliation act. Although Congressman Roberts, when debating the USDA's FY96 appropriations, had assured the opponents of farm commodity programs that there would be just such an opportunity, the risks to individuals like Congressman Roberts of floor amendments would be less if the Freedom to Farm bill could be attached to another piece of legislation such as a continuing resolution. Thus, Congressman Roberts explored opportunities to attach the proposed farm legislation to other legislation in the hopes that doing so would minimize the possibility of an adverse vote.

Senator Lugar continued to be supportive of the Freedom to Farm approach to commodity legislation. For example, in mid-December he sent a letter to the Senate Budget Committee chairman, Sen. Pete V. Domenici, R-N.Mex., and joined Congressman Roberts in sending a similar letter to President Clinton. In these letters the fixed payment approach of Freedom to Farm was endorsed.[5] However, it was not until Friday, January 25, 1996, that Senator Lugar introduced the Freedom to Farm legislation in the Senate.[6]

In the meantime, Senator Dorgan prepared to introduce legislation that would extend the then current law.[7] Congressman Roberts let it be known that he was opposed to an extension.[8] However, Senator Lugar announced that Congress should either pass Freedom to Farm as a separate bill or extend the 1990 law for one or two years.[9] Democrats, like Sens. Dale Bumpers, D-Ark., and David Pryor, D-Ark., demonstrated their strained relationship with the administration by stating that the administration should not let the Republicans shut down agriculture as they had done the government.[10]

Throughout the turmoil Congressman Roberts remained opposed to an extension of the 1990 Farm Act. With an extension, the next serious consideration of farm legislation would probably not be until 1997 after the presidential election. Although there might then be a Republican president, the accepted wisdom was that at that time the CBO baseline for budget scoring for farm commodity programs would be sharply smaller given U.S. export and farm commodity price prospects. Congressman Roberts not only coveted the early 1995 budget numbers associated with commodities, but he was also adamant that the outlays they allowed not be diverted to other purposes. He strongly opposed Secretary Glickman's suggestion to lower the transfers to crop producers in years of high prices and use that money instead for rural development or research. He stated that, "We won't touch the market transition payments. We should not be in a position to use those funds as a bank for the Secretary to use for rural development."[11] This statement would seem to mean, "Keep your hands off of the forty-some billion dollars that we want to directly benefit producers and farmland owners. If you want a few million dollars for research, credit, consumer, and commodity information programs, get it from some other place."

Republican and Democratic legislators interested in maximizing the federal transfers to crop producers and farmland owners had two major challenges that were intertwined. One was to avoid a simple extension of the 1990 Farm Act. The other was the continuing challenge to somehow "capture" the old agriculture budget base and, if that should prove to be impossible, prevent its further erosion into the spring as a result of the expected continued increases in commodity prices.

Semantics became tremendously important in meeting the challenge of capturing the commodity program baseline. The Republican leadership had repeatedly emphasized reductions in outlays. Reductions were understood to be measured from CBO estimated outlays into the future assuming no change in policy. Such estimates require assumptions about other variables. For example, CBO estimates about tax revenues and unemployment outlays are based on expected economic activities. For agricultural commodities assumptions about future exports and weather are of tremendous importance.

The difficulty for the aggies was that the prospects for agriculture commodity prices that escalated during the fall of 1995 could continue to escalate into 1996. Thus, it seemed inevitable that the lower outlay expectations associated with a continuation of current policies would eventually be officially recognized by the CBO. It would then be obvious that the $13 billion of presumed savings of Freedom to Farm versus the continuation of the 1990 act was an illusion and incompatible with a planned Freedom to Farm outlay of $43 billion. The CBO made the inevitable clear with the release of their December 1995 update. In it they indicated that they were reducing their estimate of farm program outlays for FY1996 by $5 billion and their estimate for FY1997 by $3 billion. Their estimates for the years 1998 through 2002 were unchanged from their earlier estimate.[12]

Obviously, proponents of farm commodity programs were more comfortable achieving a "calculated budget saving of $13 billion" if it were measured from the February 1995 CBO baseline of $56.6 billion than if it were measured from the end-of-1995 baseline of $48.6 billion. In one case they could embrace programs that would be estimated by the CBO to cost $43.6 billion; in the other, $35.6 billion. In practice "capturing the baseline" meant one of two things. One would be to continue to calculate savings as the difference between the February 1995 $56.6 billion and the CBO analysts' estimates of costs associated with new policies. The other would be to use end-of-1995 $48.6 billion in the calculations instead of the February 1995 $56.6 billion, but for the congressional leadership to be more relaxed about the amount of savings. Either way the bottom line for outlays would be the same. In the end the Agriculture Committees outflanked the Budget Committees and their discipline. The Agriculture Committees locked in payments to commodity producers and farmland owners that in total were nearly equal to the payments these recipients had received in the previous seven years and were somewhat less than CBO's end-of-1995 estimate of outlays with an extension of the 1990 Farm Act. This was all done while claiming budget savings of $13 billion!

Thus, a shell game ensued. The challenge was to shift the attention away from emphasizing budget savings to getting the public to affix its attention on the proposed level of spending and the fact that this level was less (by an unspecified amount) than it had been. To publicize a lower level of savings was anathema. Rather than focus on reductions of agriculture outlays of $12.3 billion (from stale and out-of-date baseline estimates) over seven years, it was politically timely to emphasize that outlays with proposed legislation would be $44 billion and that these were "transition payments" to an implied, but never stated, free market at the end of seven years. However, Congressman Roberts never claimed that Freedom to Farm would bring an end to transfers to farmers and landowners at the end of the "transition." Further, none of the major bills considered by the Agriculture Committees in the House or in the Senate included language declaring an end of farm commodity programs at any date.

The specific CBO February 1995 estimate of related farm commodity outlays in the seven years 1996–2002 was $56.6 billion. The comparable number included in the December 1995 estimates was $48.6 billion.

Thus, the so-called "erosion of the baseline" was a powerful incentive for those interested in maximizing transfers to the farm sector to avoid a simple extension of current legislation. Only a small number of legislators were pushing that option. And serious public discussion of the one-year extension option was carefully avoided. The Republicans correctly anticipated that the Democrats would avoid any posture that appeared to call openly for a smaller level of transfers. Environmentalists were also shy about advocating lower outlays. The lower the outlays available for agriculture programs, the less likely environmentalists would be able to arrange for money to be allocated for programs they favored and the less likely farmers would have strong incentives to maintain conservation compliance. To some extent environmental programs and certainly rural development programs were residual claimants.

The Interest Groups

A number of agriculturally related organizations now publicly endorsed the Freedom to Farm proposal of the reconciliation legislation that President Clinton had vetoed. For example, the National Corn Growers opposed a continuation of current programs, supported the Freedom to Farm approach included in the budget reconciliation bill, emphasized that their members needed to know the applicable 1996 policy, and advocated that caps not be placed on loan rates.[13] The American Farm Bureau Federation (AFBF) also endorsed the Freedom to Farm approach.

In mid-December the House Agriculture Committee released a list of some twenty-six organizations and groups (some of which represented several organizations and corporations) who were supporting Freedom to Farm.[14] Congressman Roberts's proposal was now getting stronger support than it had when he tried to get the House Agriculture Committee to approve it in September.

In contrast to the support by these twenty-six organizations and groups for Roberts's Freedom to Farm, the American Soybean Association expressed support for the farm security act proposed by Senator Daschle. They particularly identified the merits of the proposed marketing loan for soybeans set at 95 percent of average historical prices.[15]

Under the auspices of the Coalition for a Competitive Food and Agriculture System, the organization of agribusinesses pressing for an end to annual withholding of land from production, fifteen economists joined in signing a letter endorsing the Freedom to Farm proposal included in the budget reconciliation bill.[16] Some of the economists had been in political positions in earlier administrations, some were consultants, others were in public academic positions. Their endorsement of Freedom to Farm emphasized the effect it would have on the economic efficiency of the use of farm resources. The magnitude of the income transfers to operators and landowners, how the amount of these transfers would compare with transfers if the 1990 act was extended, and consideration of the benefits associated with Freedom to Farm as compared to benefits of using the money differently were not mentioned. Income and wealth distribution effects were not part of the letter. Aside from the endorsement, the letter particularly emphasized the importance of avoiding price supports above "long-term market clearing levels"—but didn't indicate the difficulty of ascertaining these levels on an ex ante basis.

In mid-January the AFBF again signaled their support of Freedom to Farm with their annual meeting endorsement of the legislation if it provided additional tax relief and regulatory reform as the federation had advocated in 1995.[17]

In summary, the effects of the bullish commodity markets on the farm legislation dynamics were of two major types. The buoyancy of the commodity markets was strengthening support for Freedom to Farm among operators and farmland owners, as well as agribusinesses, but simultaneously weakening the very logic undergirding federal transfers to producers and farmland owners. Congressman Roberts shrewdly anticipated both. The challenge for him was to have congressional action once there was sufficient producer, landowner, and agribusiness support, but before the logic for federal transfers was completely wiped out by the bullishness of the commodity markets.

Thus, as the farm bill debate entered its final stages in late January 1996 the focus for completing farm legislation was on the Agriculture

Committees as balanced budget activities became less demanding and the power of the congressional budget-cutters waned. Momentum was also building for Freedom to Farm as agricultural commodity and agribusiness interests began to perceive the magnitude of the transfers and the attractiveness of cropping flexibility with Congressman Roberts's proposed program.

Negotiations in the Senate 9

In the fall of 1995 during the budget reconciliation negotiations Sen. Richard G. Lugar, R-Ind., had been persuaded to support the House Agriculture Committee chairman's Freedom to Farm proposal. However, with the failure of the White House-congressional budget summit, farm legislation became separated from the budget reconciliation bill. This change meant that Senator Lugar's farm bill challenge was distinctly different than when preparing the agricultural title for the Senate budget reconciliation bill. He now needed sixty votes to overcome a filibuster threat. Thus, he could not move ahead without negotiating with the Democrats. In addition, Senate Majority Leader Robert Dole, R-Kans., continued to be circumspect about endorsing his fellow Kansan's Freedom to Farm proposal.

It was not clear until near the very end that "Freedom to Farm" would be the commodity title of the Senate bill. A proposal pressed by Senate Minority Leader Thomas A. Daschle, D-S.Dak., was seriously considered by Republican farm state senators. The irony of the eventual activity in the Senate in February 1996 is that if it had not been for Democratic Sen. Patrick J. Leahy of Vermont, the Republicans' approach of planting flexibility and seven years of guaranteed payments to producers and farmland owners might not have passed the Senate, survived the House-Senate conference, and become law.

Negotiating with Democrats

The challenge for Senator Lugar, chairman of the Senate Agriculture Committee, in the post-budget summit days was quite different than when preparing the agricultural title for the budget reconciliation bill in the fall of 1995. Rules applicable to budget reconciliation legislation differ from the rules for most other Senate legislation. Only a majority of votes, fifty-one, is required to end debate on budget reconciliation. However, for other legislation, like a stand-alone farm bill, sixty votes are required. Thus, Senator Lugar was challenged to discover the combination that would attract support of a significant number of Democrats, as well as retain the support of Republicans, especially those on the Senate Agriculture Committee, including the majority leader, Senator Dole.

Although the proposal favored by Senator Daschle was seriously considered, the winning combination included the commodity program advocated by Congressman Pat Roberts, R-Kans., in the House, but also noncommodity programs advocated by Senator Leahy, the ranking Democrat on the Senate Agriculture Committee.

The negotiations for Democratic votes had two components. One was concentrated among the senators of the Corn Belt and the Great Plains. The other and presumably later negotiation was concentrated with Senator Leahy, who was able to deliver additional Democratic votes for a farm bill if it permitted the initiation of a northeast dairy compact, reauthorized nutrition programs, continued selected environment and conservation programs, and initiated still other environmental programs.

The negotiations with the Corn Belt and Great Plains senators came close to finalizing a compromise that met Senator Daschle's requirements. However, in the end it collapsed as Republicans forced their leaders to back off from what was presumably a tentative agreement among the negotiators including the Majority Leader Dole and Senators Daschle and Lugar.

Senator Leahy, it turns out, unlike Senator Daschle, was willing to accept the Republicans' Freedom to Farm approach to commodity policy so long as his dairy compact, nutrition, and environment/conservation interests were met. An important question is why the negotiations with Senator Daschle collapsed. One sequence may have been a collective judgment on the part of the Republicans that Senator Daschle's demands in exchange for not filibustering were too high and that they would prefer to hang tough on Freedom to Farm. Commodity prices continued to be strong. They may have thought that constituent pressures supporting the fixed payment approach (regardless of commodity prices) of Freedom to Farm might nudge a sufficient number of Democrats to vote for it. And they may have expected that some of the rules applicable to rice and cotton could be made different than for corn and wheat in order to make Freedom to Farm more attractive to senators with constituents interested in these commodities.

But there is also the possibility that the negotiations collapsed because the Republicans became aware of Senator Leahy's thinking just in time to back off of any agreement with Senator Daschle. If so, they may have found Senator Daschle's requirement to support the farm bill—to modify the commodity part of the bill—more onerous than Senator Leahy's requirements—to permit the northeast dairy compact and reauthorize nutrition and environment/conservation programs, but without changes in the Freedom to Farm approach.

At any rate on February 8 the U.S. Senate with a vote of 64 yeas and 32 nays approved farm legislation that included commodity legislation similar to that advocated by Congressman Roberts in the House.

A Chronicle

Several events reported by news people between January 23 and Saturday, February 10 provide a sense of the farm legislation drama unfolding in the Senate offices during this short period of time. These events illustrate (1) that Senate action on farm legislation was very unpredictable until near the very end, (2) the way that the sixty-vote cloture rule forces compromise and logrolling, and (3) how the timing and content of Senator Leahy's approach to the Republicans had an overwhelming effect on the eventual Senate-approved farm bill and then later on the legislation passed by the Congress and signed by President Clinton.

Tuesday, January 23
Sen. Byron L. Dorgan, D-N.Dak., introduces S. 1523, which would extend the current farm legislation.[1]

Senator Dole publicly acknowledges that talks about farm legislation among staff members will take place.[2]

Senator Dole states, "We ought to go back and make some changes in the 'Freedom to Farm Act' ... so that it will have enough support to pass both the House and the Senate."[3]

Friday, January 26
Senator Lugar and fourteen others including Sens. Robert Dole, R-Kans., Thad Cochran, R-Miss., and Larry E. Craig, R-Idaho, introduce S. 1541 in the Senate. It is very close to the farm legislation that had been included in the budget reconciliation bill vetoed by President Clinton.

Monday, January 29
Senator Lugar acknowledges that it is very difficult to reach agree-

ment on a combination of provisions that could attract the sixty votes. He attributes the difficulty to a polarization of views among Republicans and among Democrats.[4]

Senator Dole schedules a floor debate on the farm bill for Thursday, February 1.[5]

Tuesday, January 30
Senator Lugar indicates that the Senate will have three choices when it votes on farm legislation—a one-year extension of the 1990 Farm Act, the Republican Freedom to Farm proposal, and a Senator Leahy compromise.[6]

Senator Craig indicates that Republicans are open to an agreement that would pay less when crop prices are high, but keep it in reserve for when prices fall or disaster hits.[7] Senator Craig is quoted as saying, "I think everything is on the table at this moment. We recognize that we've got to craft a bipartisan compromise."

Senator Daschle reports that staff aides are working out a bipartisan farm bill and that he is hopeful that a compromise will be ready by Thursday, February 1. He visualizes that the sequence of voting will be first on a motion to limit debate on the Republican bill; second on a Democratic compromise; and third, "We'll get to the compromise, should it be agreed on."[8]

Wednesday, January 31
Senator Leahy indicates that he will support Freedom to Farm if nutrition and conservation are "shielded" and the northeast dairy compact is included in the prospective bill. On that day Senator Daschle is out of town, and he tells reporters that he was still working with Senator Dole to develop a compromise.

The Clinton administration's OMB sends a letter to the Senate leadership regarding farm legislation. It states that the president would veto the bill being considered because of its effects on the "safety net." The OMB letter makes no mention of OMB attitude on the provisions subsequently added at the behest of Senator Leahy.[9]

Thursday, February 1
The Senate is scheduled to vote at 1:30 P.M. on whether to end debate on a farm bill sponsored by Senator Dorgan. The bill would extend, with a limited number of changes, the 1990 farm legislation. If that is

defeated, the Senate will vote on a compromise worked out among Senators Leahy, Lugar, and Dole. The compromise is expected to include Freedom to Farm commodity legislation, but in addition legislation on other topics requested by Senator Leahy and agreed to on January 31.[10]

Senator Dole delays cloture vote on the Republican farm bill after concluding that there are not the necessary sixty votes to approve cloture.[11]

Cloture motion in the Senate to limit debate on S. 1541 fails with a vote of 53 yeas and 45 nays.[12]

Senator Daschle indicates that, "We can't get an agreement if there's an insistence on 100 percent payment. That's not going to work." Senator Leahy's staff indicates that an agreement is certain to happen, and Senator Lugar tells the press that they are close to agreement.[13]

Senate staffers describe major features reportedly agreed to by Senate leaders for a bipartisan farm bill. The features included: reauthorization of nutrition programs for seven years; authorization of crop supports for three years, not seven as was proposed by the House; and Senator Daschle's proposed annual nonrefundable advance payments each year equal to 40 percent of a payment based on historical subsidies, with the other 60 percent linked to market prices and county yields. In addition, producers would have wide planting flexibility, the CRP would be capped at 36.4 million acres, new CRP enrollments would be allowed, an EQIP (developed in the summer of 1995 by the Republican and Democratic Senate Agriculture Committee staffs) would be implemented, crop insurance would be mandatory, loan rates would not be capped, and annual land retirement would end.[14]

Senator Dole states on the floor that a number of members, both Democrats and Republicans, had been working on the farm bill for the "last two to three hours." He adds that there may not yet be agreement but that they may be "close to an agreement."[15]

Senator Dole says that the Senate will vote on cloture on Thursday evening if the votes are there for cloture. If not the farm bill will be taken up on Friday or the following week.[16]

Shortly before the time that Senator Dole was expected to announce a farm bill cloture vote, he announces on the floor that it will be "temporarily" delayed. In the same Dow Jones news release reference is made to new language developed by Senator Leahy and agreed to with Senators Lugar and Dole on Wednesday, January 31.[17]

Senator Dole reports that efforts to design a farm bill have failed and that further efforts will be delayed until the next week.[18]

Friday, February 2
Secretary of Agriculture Glickman indicates that he would recommend a presidential veto of the present draft of the compromise bill.

Senator Lugar describes what he now visualizes as a bipartisan farm bill. The guaranteed payment would be 40 to 50 percent of a payment estimated with the remainder linked to market conditions. Further, he expects the bill to be three to five years instead of the seven years contemplated earlier. He reports that Secretary Glickman's proposal of a rural development fund was controversial but that it might be part of a bipartisan bill.[19]

Secretary of Agriculture Glickman indicates that the discussions among Senate and USDA staffers are moving in the right direction. He emphasizes a safety net (which is interpreted as meaning payments to farmers are linked to market conditions), the desirability of planting flexibility, his requirements for a rural development fund, allowing reenrollments in a reauthorized CRP, and planting flexibility. Greg Frazier, Glickman's chief of staff, criticizes the House dairy plan that would require the additions of dry milk solids to fluid milk for drinking and thereby raise milk prices and the cost of public feeding programs.[20]

The *CBS Evening News* reports that the dairy provision of the House bill would raise the retail price of milk about forty cents a gallon.[21] The issue catches public attention.

Monday, February 5
Fourteen Republican senators from farm states say that they will not support the compromise worked out with Senator Daschle.

Tuesday, February 6
Senator Dorgan makes reference to a compromise and states that, "We had a compromise...over the weekend...Yet we are told by some, 'Either you invoke cloture and cut off debate and cut off amendments on the freedom-to-farm bill or we are not going to play....'"

Senate cloture vote on farm legislation compromise developed by Senators Lugar and Leahy is scheduled for midday February 6.[23]

Sixteen Republican senators move "to bring to a close debate"—cloture—on Amendment No. 1384 to Senate bill 1541 (technically on Senator Leahy's bill). The vote fails by one vote with 59 yeas and 34 nays. In response to the vote Senator Dole states, "I will let the Democratic leader know whether we will have another cloture vote on Thursday. But I think it is pretty obvious that had our absentees been here, we would have had cloture, and have pretty good bipartisan support. It seems to me that we are pretty close to a bipartisan resolution of this matter."[24]

Sens. Trent Lott, R-Miss., Phil Gramm, R-Tex., and three other Republicans are not present for the cloture vote. Eleven Democrats vote for cloture making it evident to Senator Daschle that the battle is over.[25]

When the cloture vote fails by only one vote Senator Daschle acquiesces to having a vote on the Senator Leahy-supported bill.

Wednesday, February 7
Sparks Commodities reports that negotiations to develop a compromise that had continued over the previous weekend had broken down and accounted for "Tuesday's unsuccessful vote to get a measure to the Senate floor" on that day, February 6. Senator Daschle is reported to be protesting draconian cuts contained in the Lugar and Leahy bill, and Republicans claim that the negotiations failed because the Democrats did not get enough Democrats to sign onto the compromise.[26]

The Senate rejects a major amendment (Amendment 3452) offered by Senator Daschle by a vote of 63 nays and 33 yeas. The Daschle amendment would have made the farm bill a three-year rather than a seven-year bill, and provided guaranteed payments to producers and landowners at 40 percent of the Freedom to Farm level, safety net payments geared to market prices and county yields, marketing loans at 90 percent of the five-year olympic average of prices, and a continuation of 1938 and 1949 law as "permanent law."[27]

Previous to the vote on Amendment No. 3452, Sen. Orrin G. Hatch, R-Utah, withdraws an amendment that would have permitted the use of food stamps to purchase "nutritional supplements of vitamins, minerals, or vitamins and minerals" after reaching agreement with Senators

Lugar and Leahy that hearings will be held regarding the proposed change.[28]

Previous to the vote on Senator Daschle's amendment, the Senate votes on Amendment No. 3451 proposed by Senator Dorgan. The amendment would have required planting of base acreage to program crops in order to receive the Freedom to Farm payments.[29] The amendment was rejected by a vote of 48 yeas and 48 nays.[30]

Along with several other amendments the Senate gives unanimous consent to including Amendment 3456 in S. 1541, the Senator Leahy/Senator Lugar compromise. Amendment No. 3456 calls for the farm price support of rice to be set between 50 percent and 90 percent of parity "as the Secretary determines will not result in increasing stocks of rice to the Commodity Credit Corporation."[31]

Near the end of the Senate floor consideration of the farm legislation Senator Daschle and his colleagues are able to obtain agreement that at the end of 2002, upon expiration of the proposed legislation the 1949 farm law would become effective again—unless a new law is passed. If this were part of the prospective new farm act, Congress would be forced to consider farm legislation rather than let farm commodity programs quietly expire. This effort by Senator Daschle reflects his strong sentiments and those of several of his colleagues, especially those from in the northern plains states, that transfers to producers and farmland owners should continue beyond 2002.

Amendment No. 3184 to S. 1541, the Senator Leahy/Senator Lugar compromise, passes the Senate this time by a vote of 64 yeas and 32 nays. Twenty Democrats join forty-four Republicans in voting for S. 3184.[32]

By a vote of 64 yeas (including twenty Democrats) and 32 nays the Senate passes S. 1541 as amended. Sen. Rick Santorum, R-Pa., moves an amendment to reduce peanut price supports; it is defeated. Senator Dole moves that the amendment be tabled; it is.[33]

The Senate thus passes a farm bill before the Senate recess that includes commodity programs very similar to those approved by the House Agriculture Committee. However, it omits a dairy title. But the bill includes trade, research, conservation, credit, nutrition, and rural development titles that were not then in the bill reported out of the

House Agriculture Committee and ready to be voted on by the full House.³⁴

After the vote on Wednesday Senator Dole states that, "I believe we need to make certain this is going to work so that we don't have these stories appearing that somebody had a big crop and got a big payment."³⁵

Friday, February 9
President Clinton indicates that he supports the freedom that the Senate farm bill would give farmers, but that he is concerned that it would erode support for farm programs by the public and could lead to the United States discontinuing its subsidies programs for farmers without other countries doing likewise. He also indicates that he did not favor fixed payments, that payments should be linked to prices, and that to get a payment, recipients should farm. But he does not threaten a veto.³⁶

Saturday, February 10
On this day a Senate aide is reported to have said that Senator Leahy came to them with ten votes if the bill included measures that Leahy desired on nutrition, conservation, and the northeast dairy compact. Ed Barron, minority chief of staff for the Senate Agriculture Committee, is quoted as saying, "Sen. Leahy's goal was to work out a comprehensive, bipartisan farm bill that also protects the environment and nutrition programs." Nancy Danielson of the Farmers Union points out that Senator Leahy broke ranks with the other Democrats. Previously they were working together, and his breaking ranks changed the dynamics.³⁷

President Clinton in his radio broadcast repeats several of the points made the previous day. He voices concern about (1) the Senate bill not including safety net provisions that would assist farmers in low-price years, (2) the expectation that people would receive payments even if they did not grow a crop, and (3) doing away with farm programs even before other countries eliminated their programs.³⁸

THE END OF SENATE NEGOTIATIONS

Observers of the many diverse opportunities leading to the climactic vote in the Senate can't help but contemplate possible alternative outcomes. For example, what might have happened in the Senate if the administration had more favorably considered Senator Daschle's forty-sixty proposal and then worked hard to achieve its passage? And what if Senator Leahy's and Senator Daschle's staffs had worked more closely together and the senators together insisted on a package that did not include Freedom to Farm-type legislation? There was a moment where things might have gone either way. The Senate might have passed a Daschle-type bill, and such a development might have affected the administration's activities and influenced what the House passed, in turn influencing what was agreed to in conference between the Senate and House. This was a deciding moment on the path to the 1996 Farm Act.

House Approval of Freedom to Farm 10

Once it became obvious that the White House-congressional budget summit was to come to naught in terms of an agreed-on a budget reconciliation bill, farm bill activities in the House became more intense, particularly in the latter part of January 1996. Congressman Pat Roberts, R-Kans., pressed for passage of his proposed Freedom to Farm legislation. The Democrats continued to experience difficulty in developing a definitive proposal that would attract credible support.

In the end, Congressman Roberts's tenacity along with the impact of the bullish farm commodity markets prevailed. On February 29 the House approved Freedom to Farm by a vote of 270 yeas and 155 nays.

Congressman Roberts's Attachment to Freedom to Farm

In the post-budget summit days Congressman Roberts intensified his efforts to attain the passage of Freedom to Farm. It was the central part of H.R. 2854 that he had introduced in the House in early January 1996. In his pursuit to find a way to get passage of Freedom to Farm he refused to move in any way from the market transition payments, the central feature of the legislation. And he resisted all attempts to reduce market transition payments by even just small amounts in order to fund other rural activities such as those related to rural development. Consistent with his strong advocacy of Freedom to Farm, Congressman Roberts was adamantly opposed to an extension of the 1990 farm legislation.[1]

Congressman Roberts's situation in late January differed in a major way from what it had been in the fall of 1995 when preparing the agriculture component for the budget reconciliation bill. In the fall the House leadership was willing to take extraordinary steps such as inserting Freedom to Farm into the House budget reconciliation bill even though it was not voted out of the Agriculture Committee. These types of end runs were no longer possible in the post-budget summit weeks.

In response Congressman Roberts adjusted his approach while at the same time retaining his attachment to Freedom to Farm. During the latter part of January, as part of his efforts to increase the support for his Freedom to Farm proposal, he asked the House leadership to commit an additional $2.5 billion for rural development activities that were favored by the Democrats. The additional $2.5 billion would have made it easier for House Democrats to support a bill containing Freedom to Farm. However, this request for funds was denied by the House leadership. Congressman Roberts was unwilling to scale back the prospective outlays for commodities in order to allocate $2.5 billion for such purposes.[2] Another example of how he adjusted his approach in order to achieve House approval of Freedom to Farm was his acceptance of the inclusion of a conservation/environment title in the bill that finally passed the House.

At the same time, an increasing proportion of farm program participants concluded that Freedom to Farm would mean more money for them. They told their congressional delegations—Republican and Democrat alike—about their support of Freedom to Farm. True, the bullish market conditions made the "continuation of the 1990 Farm Act" option increasingly attractive to balanced budget advocates. But even the really strong balanced budget advocates had no fire in their belly to take on the aggies at this stage.

In addition, and not surprisingly, the attitudes of Congressman Roberts's fellow Republicans on the House Agriculture Committee were markedly different in early 1996 from the situation in November 1995. In November some fellow Republicans by their no votes made it impossible for Congressman Roberts to achieve a favorable vote on his proposed farm bill for inclusion in the budget reconciliation legislation. But in late January 1996, all of the Republicans on the House Agriculture Committee, including those with cotton and rice constituents, were backing Congressman Roberts's proposal and his efforts to gain its passage in the House. In caucus they endorsed his efforts to attach the farm bill to "must pass" type legislation such as the bill designed to avoid another shutdown of the government and the one that would raise the U.S. government debt limit.

Democratic Reactions

House Democrats were largely in a reactive mode, as was the administration, even though several proposals swirled around as prospective markups of agriculture legislation approached. Finally, even Congressman Charles W. Stenholm, D-Tex., suggested that the Congress pass a Freedom to Farm-type bill, but for only two years.[3]

At the suggestion of the secretary of agriculture, Republican and Democratic Agriculture Committee members of the House and the Senate met with the secretary on January 23. This meeting was just previous to the secretary's departure for a ten-day trip to Asia. The meeting was billed by some observers as a bipartisan summit. The discussion touched on the many alternative courses of action that might be pursued. However, it obviously did not generate agreement among the participants.[4]

ANOTHER TRY AT DAIRY

Amid the flurry of activity to bring closure to the preparations for the House Agriculture Committee markup of the farm bill, Congressman Steve Gunderson, R-Wis., worked feverishly to develop yet another dairy bill in time for it to be voted on in a coming Agriculture Committee markup session. This time Congressman Gunderson embraced a proposal developed largely by the National Milk Producers Federation. It was strongly opposed immediately by the International Dairy Foods Association.

Congressman Gunderson's proposed dairy title called for

—reducing the number of milk marketing orders from the present thirty-three to a number between eight and thirteen,
—requiring the addition of milk nonfat solids to milk sold for fluid consumption to meet the levels now required in California, and
—discontinuing the federal government's purchase of butter and powdered milk as a means of supporting farm milk prices, but continuing to support milk prices with the purchase of cheese. Initially the milk price support price would be $10.35 per hundredweight, its current support price. Starting in 1995 this support price would be lowered by ten cents each year for five years.[5]

Opponents of these provisions organized an attack focusing on the proposed required additional milk nonfat solids. They correctly recognized that this was an Achilles' heel of the Gunderson proposal. For example, a USDA analysis concluded that the higher requirements for the nonfat solids standards combined with the higher minimum Class I prices would raise the price of milk by thirteen cents per hundredweight for the years 1996–2002. The effect on farm income would mean an increase of $3.4 billion for the seven years. Dairy program costs would decline by $625 million. However, consumer costs would increase ten cents to thirteen cents per gallon for whole and skim milk and by twenty-two cents to twenty-four cents for low-fat milk. Child nutrition and food stamp program costs would rise $1 billion over the seven years FY1996–FY2002, and the cost of the WIC program would increase $130 million if participation continued at present levels.[6] Opponents, including some in the dairy industry, encouraged media attention to the potential increase in consumer costs implicit in the proposal. This attention contributed immensely to the demise of the proposal.

House Agriculture Committee Markup

With the inability to attach Freedom to Farm legislation to a piece of "must" legislation and the February congressional recess approaching, Agriculture Committee Chairman Roberts did the obvious. Knowing that all of the Republicans on his committee would vote for Freedom to Farm, he scheduled a committee markup.

On January 30 the House Agriculture Committee met to consider freestanding commodity legislation. In the end the committee, by a vote of 27 to 19, approved Congressman Roberts's proposed commodity legislation calling for outlays of $43.6 billion over seven years. Three Democrats voted with the twenty-four Republicans on the committee.[7]

The three Democrats voting with the Republicans were Congressmen Collin C. Peterson, D-Minn., Gary A. Condit, D-Calif., and Sanford D. Bishop Jr., D-Ga. Congressman Peterson explained his affirmative vote as a way to be involved in negotiations with the Senate and because of the importance of dairying to Minnesota. Congressman Bishop's support was related to the inclusion of changes in the peanut program in a bill that had support of the Republican leadership. He reasoned that the committee-passed bill provided shelter for the peanut program from more drastic reforms. Reports on Congressman Condit's rationale were more sketchy.[8]

Early in the markup session Congressman Calvin M. Dooley, D-Calif., introduced an en bloc amendment that, if approved, would have cut the proposed fixed annual transition payments in half; established marketing loan rates for wheat, feed grain, cotton, rice, and oilseeds at 90 percent of the olympic average of the past five years of prices for these commodities; and provided $7.5 billion for rural development, crop insurance, and research. It was defeated by a vote of 18 to 27.[9]

One amendment offered by a Democrat was accepted by the committee. All others were defeated. The accepted amendment was proposed by Congressman Peterson. It called for capping the CRP at 36.4

acres and eliminating a provision that would have allowed owners of land enrolled in the CRP to terminate the CRP contract on that land by giving a sixty-day notice.[10] The amendment was viewed favorably by conservation-oriented groups and was consistent with a draft conservation title developed much earlier by Republican and Democratic staff of the Senate Agriculture Committee.

Congressman Roberts had the support of all of his Republican colleagues and was steadfast in defending the entirety of his Freedom to Farm title and the dairy title sponsored by Congressman Gunderson. Throughout the markup session the Republicans directly defeated eight amendments offered by the Democrats. In frustration the Democrats decided to forgo offering any more amendments knowing that they would surely be defeated even though additional amendments had been prepared.

Two Goals

Subsequent to the committee markup Congressman Roberts immediately pursued two tactical objectives. One was to obtain a "closed rule" so that at most one substitute amendment could be offered on the floor. The other objective was to have the bill come to the floor previous to the House recess, which was to start on February 2. He was unable to achieve either objective. At the same time that Congressman Roberts was pressing for an early vote, Speaker Newt Gingrich, R-Ga., was saying that a vote on the farm bill might not occur until the end of February.[11] Perhaps the Speaker realized that a rule on the bill from the House Rules Committee chaired by Gerald B. H. Solomon, R-N.Y., would not be forthcoming before the House recess. Congressman Solomon was not happy with Congressman Gunderson's dairy title.

There was another reason why Congressman Roberts did not get a closed rule or a vote before the recess. The Democrats took the full technically available time to file their dissenting views for inclusion in the committee report. House procedures call for committee reports to be available previous to a House vote. Although as expected Roberts berated the Democrats for their tactics, it is a bit surprising that he had not scheduled the committee markup for an earlier date. An earlier markup would have preempted the possibility of the Democrats delaying the vote through taking the maximum time to file their dissents.

As it turned out, the delay of the conference report was probably inconsequential. The eventual floor debate revealed that Congressman Solomon was strongly opposed to the dairy title of the committee-passed bill. Thus, the opportunity for a closed rule was probably nil from the very start. In addition, Congressman Solomon's staff probably needed some time to formulate his substitute dairy provisions to offer as an amendment once the marked-up bill was on the floor.

The House recessed without floor action on the farm bill marked up by the House Agriculture Committee.

In the House in Late February 1996

When Congress returned from recess on February 26, the Senate had passed its bill and those with financial or political stakes in farm legislation focused on House activities. The farm bill that had been voted out of the House Agriculture Committee was near the top of the list for floor action. However, two considerations had to be dealt with before H.R. 2854 would go to the floor and be voted upon by the House of Representatives. One was a jurisdictional question between the House Appropriations Committee and the House Agriculture Committee. The other consideration was whether the bill would be subject to amendments on the floor and, if so, how many. The first was dealt with quickly and with great effect on the Agriculture Committee activities. The second dealt with the scope of the legislation and therefore was fundamental. It was not actually settled until voting on amendments was completed.

Turf Again

During the recess the feud between the House Appropriations Committee and the Agriculture Committee arose again. On or about February 20 Congressman Bob Livingston, R-La., chairman of the Appropriations Committee, and Congressman Joe Skeen, R-N.Mex., chairman of the Agriculture Appropriations Subcommittee, wrote to House Speaker Gingrich and argued that the House of Representatives' proposed farm legislation, as well as the farm bill passed by the Senate, violated the jurisdiction of the House Appropriations Committee. In addition, they argued, the bills included new spending that was inconsistent with the goal of balancing the budget by 2002.[12]

Although the challenge would be a harbinger of future difficulties for the Agriculture Committee, this time it was mostly a nuisance.

Scope

Congressman Roberts had asked for a closed rule on H.R. 2854 whereby no amendments would be allowed on the floor. The House leadership, however, decided that the legislation should be considered under a modified rule whereby a limited number of amendments could be submitted during floor debate. Thus, the House Rules Committee was charged with deciding which proposed amendments to accept. The Rules Committee was chaired by Congressman Solomon, who had a strong interest in dairy legislation.

Congressman Roberts was sensitive to the Senate approval of a bill that with Sen. Patrick J. Leahy's, D-Vt., insistence and Sen. Richard G. Lugar's, R-Ind., acquiescence (if not support) included several noncommodity titles. In spite of the necessity for the Senate to move toward this broader design in order to attract sufficient votes to avoid a filibuster, Congressman Roberts continued to publicly reject such an approach. For Congressman Roberts, keeping the legislation separate and concluding commodity legislation first by itself minimized the chances that others would attempt to lower the proposed commodity-related payments to individual recipients in order to finance federal support for credit, rural development, research, and conservation. Congressmen Roberts and Gunderson, in criticizing the Senate bill for including titles other than commodities, pointed out that the noncommodity provisions in the Senate bill were adding $2.4 billion in mandatory spending and $735 million in discretionary spending.[13] The Senate, on the other hand, realized that once Congressman Roberts had obtained congressional approval of his Freedom to Farm there would be no way to leverage him to accept the noncommodity titles.

These noncommodity-type programs can be very important to rural residents, including producers and farmland owners. But in many cases such benefits are indirect. There is little question but that producers and farmland owners prefer to receive the checks directly as opposed to money going to institutions with the benefits to these commodity program beneficiaries being indirect. Thus, Congressman Roberts's strategy of locking up the commodity money first and letting the supporters of the other programs scramble for any remaining money was very enticing for those with commodity interests.

Congressman Roberts argued his position publicly and, undoubtedly, privately. He promoted the concept of finalizing commodity legislation in one bill and then later in 1996, after hearings, developing a second farm bill as a vehicle for trade, conservation, research, credit, and other farm-related legislation. One step in making his allusions to a second farm bill credible was his Tuesday, February 27 introduction of H.R. 2973. The bill included titles on research, credit, conservation, and trade.

The timing of the introduction of H.R. 2973 happened to coincide with the secretary of agriculture sending a letter to Congressman Roberts. In the letter Secretary Glickman argued that the House bill should include titles related to rural development, research, trade, and conservation. He also repeated his belief that the prospective legislation should include a safety net and that contract payments should not be made if the land is not used to plant a crop or if an appropriate conservation plan is not implemented on the land.[14]

As of Tuesday morning, February 27 the House Rules Committee had received seventy-one proposed amendments to the bill reported out by the House Agriculture Committee. By that time Congressman Roberts had accepted the reality that a closed rule was no longer possible and indicated that he was hoping for no more than ten amendments being allowed by the Rules Committee.[15] However, the House Rules Committee decided to allow sixteen amendments.

On the Floor

Floor action on H.R. 2854 started on Wednesday, February 28 and was concluded the following day. The 244 yeas to 168 nays vote to call the bill for debate was a clear indication of the eventual outcome of the debate—passage of the bill as amended with 270 yeas and 155 nays. Only nineteen Republicans voted against the bill; fifty-four Democrats voted for the bill.

In total, Congressman Roberts and his colleagues were extremely successful during the floor debate. There were three disappointments, however. The biggest disappointment was the substitution of a dairy amendment sponsored by Congressman Solomon for Congressman Gunderson's compromise dairy program that Congressman Roberts had accepted as part of the committee bill, H.R. 2854. The other two disappointments were more of an embarrassment than a real loss. They were the very narrow margins by which the peanut and sugar programs were sustained into the future—the peanut program by a margin of three votes and the sugar program by a margin of nine. The sugar program may only have been maintained through whipping up strong anti-Cuban sentiment after two "private" planes from the United States were shot down over Cuba.[16]

More specifically, by the end of the first day of debate, February 28, the House had accepted by a voice vote the en bloc amendments proposed by Congressman Roberts. And the House had rejected a series of other amendments[17] that would have

—eliminated marketing loans for cotton in 1999 (sponsored by Joseph P. Kennedy II, D-Mass.; the vote was 167 yeas and 253 nays),

—phased out farm programs over five years, but retained the 1949 law (sponsored by Barney Frank, D-Mass.; the vote was by voice), and

—phased out the peanut program over seven years and lowered supports from $610 to $310 per ton in 2001 (sponsored by Christopher Shays, R-Conn., and Nita M. Lowey, D-N.Y.; the vote was 209 yeas, 212 nays).

On the following day, Thursday, February 29, the House completed action on H.R. 2854.[18] Over the objections of the chairman of the House Appropriations Committee, Congressman Livingston, the conservation amendment offered by Congressman Sherwood L. Boehlert, R-N.Y., was approved by a vote of 372 yeas and 37 nays. Congressman Boehlert's amendment called for capping the CRP at 36.4 million acres, allowing termination of the CRP contracts by CRP participants, and providing $200 million for EQIP. The House also approved, again over the objections of Congressman Livingston but with the support of Speaker Newt Gingrich, the use of $200 million for the purchase of land in the Florida Everglades with a vote of 299 yeas and 124 nays. This proposed purchase of land was related to the concern about sugar production in Florida having an adverse effect on the environment.

Although Congressman Roberts had opposed including environmental and conservation provisions in H.R. 2854 and instead favored considering it later, by the time of the floor vote he recognized that his was not a sustainable position. He then spoke favorably on these matters and, in fact, voted for both of the environmental-related amendments.

With a favorable voice vote, an international trade title dealing with, among other topics, international food assistance was added.

As mentioned above, the House approved the substitution of a dairy title proposed by Congressmen Solomon and Dooley. It was a substitute for the dairy title developed by Congressman Gunderson and embraced by Congressman Roberts. By a vote of 258 to 164 the House voted approval of the Solomon dairy plan. The news stories that focused on consumer costs and the spillover effects of the Gunderson-developed dairy title on federal outlays for food programs undoubtedly influenced the vote on the Solomon amendment.

Other proposed amendments were defeated. One was an en bloc amendment—the Democrats' substitute bill. It was introduced by E. de la Garza, D-Tex., and would have among other things created a $3.5 billion fund focused on investments in rural America. It was defeated by a vote of 253 to 167. In addition, Congressman Steve Chabot's, R-Ohio, amendment to end the cotton marketing loans was defeated 299 to 121, and Congressman Charles E. Schumer's, D-N.Y., amendment to kill the sugar program was defeated 217 to 208.

In the end, the amended bill was approved with a vote of 270 yeas and 155 nays. Congressman Roberts had skillfully protected his Freedom to Farm title and its all-important payments to producers and farmland owners—a title nearly identical to that approved in early February by the Senate. At this point, congressional approval was virtually certain, and barring a veto, producers and farmland owners would be free to use—with very modest limitations—their resources as they thought appropriate and to receive federal checks for seven years with only two criteria: that they had received commodity program checks sometime in the past five years, and that they follow conservation practices that practically all recognized as very modest requirements. The legislation that was initiated in response to a Republican drive to decrease outlays was about to become the law of the land. This was the case even though all those who understood farm programs recognized that the outlay savings with the prospective farm act might be quite modest and could even lead to higher outlays (if commodity prices were to stay high) than if the 1990 Farm Act were simply extended.

AT LAST, A FARM COMMODITY ACT   11

Three formal steps were yet to be completed before there would be a 1996 Farm Act: (1) House-Senate conferencing of the passed bills, (2) Senate and House passages of the bill agreed to in conference, and (3) signature by the president.

In conference, the Sen. Richard G. Lugar, R-Ind.,-Sen. Patrick J. Leahy, D-Vt., logroll endured. It was clear that Senator Leahy was supporting the Freedom to Farm approach to commodity policy. Leahy did not leave anything to chance, however, as to what he expected in exchange. At the very beginning of the conference, without equivocation, Senator Leahy announced that neither he nor Senator Lugar would agree to a conference agreement if it did not provide for consent with the proposed northeast interstate dairy compact. In a move to assure that all of the conferees understood his requirements on the northeast interstate dairy compact, and to demonstrate that Senator Lugar and he were of one mind on this point, Senator Leahy distributed a printed release which stated the conditions upon which he and Senator Lugar had agreed.[1]

THE NATURE OF CONFERENCING

The Senate-House conference on the farm bill legislation was held in the Senate Agriculture Committee meeting room, 301 Russell Senate Building. The room is dominated by three ornate chandeliers and a large mirror encased in twelve-inch-wide gold leaf framing hung above a marble fireplace located at one end of the room. As you enter the room the mirror is on the far end; the windows are on your right. On one side of the room there are three twenty-foot windows with nondescript dusty drapes. Around the room, at least ten feet above the floor, hang photos of past Senate Agriculture Committee chairmen. There is one portrait. It is of Herman Talmadge, the long-time chairman of the Senate committee. The portrait of Senator Leahy commissioned by Senator Lugar upon his replacing Senator Leahy as chairman was not yet hung at the time of the conference.

The conferees sit at a long rectangular table, two leather chairs on each end and twelve on each side with Senator Lugar in the center on one side and Congressman Pat Roberts, R-Kans., in the center on the other. The senators sit on the side of the table next to the windows and on the table end facing the mirror, the representatives sit on the side facing the windows and under the mirror. Behind the conferees on each side of the table are two rows of chairs for the aides. The remainder of the room, perhaps one-fourth, is crowded with folding chairs. The majority of these chairs are reserved for the press. About six chairs are reserved for USDA officials. The remaining five are for lobbyists and others who claim their chairs on a first-come-first-served basis, unless they are from a big law firm. These firms hire stand-ins to be their proxies in line. That way the first served are those who hired stand-ins who get in line early.

For the most part the discourse in conference is very cordial. The interactions among members are certainly more cordial than the discourse, especially after the completion of the formal conference sessions, among the staffs of the two committees and the aides to the conference members! By the time the conference was called to order and Congressman Roberts asked for unanimous approval for Senator Lugar to serve as chairman (to which all agreed), the staffs had pored over five hundred-some differences between the Senate-passed and the House-passed bills. They had reached tentative agreement on all but eighty-four by Sunday evening, March 17. By the start of the conference on Wednesday the unresolved list was less than twenty. So the first order of business after Senator Leahy informed the conferees about his and Senator Lugar's agreement about the northeast interstate dairy compact was to make the staff agreements official. The conferees formally approved them en bloc. As the meeting proceeded the conferees progressively whittled down the number of issues on which they disagreed, first one, then another. When agreement seemed elusive, discussion was postponed or referred to a subgroup to report back to the entire group. There was vigorous search for compromise language. Once in a while there was no alternative but to take a vote, first among the members of one side and then the other. But priority was given to searching for language where each side gave some but not all and consensus could be proclaimed by the chair.

The entire process is a grand activity of collecting and sifting information, assessing how alternative decisions will affect the public good and the welfare of particular groups and those known by the conferees, and ultimately how the outcomes will in the end affect the political fortunes of those involved.

AT LAST, A FARM COMMODITY ACT   117

The staffs are tremendously important to the entire activity. They gather, distill, and channel information from constituents, interest groups, and other staff members to their principals and vice versa.

DIFFICULT ISSUES

The List

By the start of the conference on Wednesday, March 20, 1996, the issues still to be resolved were less than twenty including among others the following:

- —Whether to reauthorize the federal food stamp program as included in the Senate-approved bill—there was agreement to do so.
- —Whether to establish a Rural Development Fund as included in the Senate-approved bill—there was agreement to do so.
- —Whether to consent to the northeast interstate dairy compact as the Senate-approved bill did—there was agreement to do so.
- —Whether to require that land for which production flexibility contract payments are received must be used for agricultural and related activities as included in the House-approved bill—there was agreement to do so.
- —Whether to establish a Commission on 21st Century Production Agriculture as included in the House-approved bill—there was agreement to do so.
- —Whether to allow CRP participants to terminate their contracts with USDA by giving the secretary a sixty-day notice as included in the House-approved bill—there was agreement to do so.
- —Whether size limits are specified for particular livestock operations participating in EQIP as included in the Senate-approved bill—there was agreement to place limits on the size of large livestock operations receiving benefits. However, the secretary of agriculture was charged with defining the meaning of large.

Three other issues were particularly contentious. They related to permanent legislation, dairy, and peanuts.

Permanent Legislation

The House-approved bill provided for the repeal of the Agricultural Act of 1949 and unused sections of the Agricultural Adjustment Act of

1938. In contrast, the Senate-approved bill did not repeal either the Agricultural Act of 1949 nor the unused sections of the Agricultural Adjustment Act of 1938. Instead, it maintained both as permanent law and provided for their temporary suspension as had been done in the past with a series of successive farm acts.

This issue was certainly the most contentious issue among the conferees. Democratic senators like Sen. Tom Harkin, D-Iowa, and Sen. Kent Conrad, D-N.Dak., argued vigorously to suspend but nonetheless retain the 1949 act and the unused sections of the 1938 act. Congressman Roberts and Senator Lugar took the opposite position.

Congressman Roberts pressed for something other than "the fishhook [the 1949 act as permanent legislation] that you had to swallow," something that was more consistent with conditions in the 1990s. He emphasized that if farming gets into big trouble, like it did in the 1980s, "We will do something; we will not jump onto the shoals of no program."

During the Wednesday debate on this issue, Senator Lugar indicated that his view was "identical" to Congressman Roberts's, even though the Senate bill contained language that merely suspended the permanent law while the House bill included language that repealed it. He reminded the conferees that the secretary had stated that it would be a catastrophe for the 1949 act to be implemented at the present time, and he said that he would "offer to recede to the House" language.

In opposition, Senator Conrad argued that without permanent law there would be pressures on the president to veto the bill. He openly recognized that keeping the 1949 act as permanent law would give him and others leverage in the future to obtain legislation they desired. Senator Leahy also argued for keeping what was the bipartisan Senate compromise—suspend related sections of the Agricultural Adjustment Act of 1938 and the Agricultural Act of 1949.

Congressman Roberts recognized that no one knew the future and stated that it was not a Republican plan to go to no program, at which point he indicated that he would think about alternative language.

As he had indicated, Senator Lugar offered a motion to repeal the 1949 act. It failed among the Senate conferees by a vote of 2 yeas to 9 nays. Sen. Thad Cochran, R-Miss., voted with Senator Lugar.

The following day Congressman Roberts offered in sequence three alternatives to merely suspending permanent legislation. One alternative was to extend the prospective legislation through the 2003 crop year. Another alternative would have retained the 1949 act through 2003 but ended it at that time. The third alternative would provide price-support loans in 2003 and beyond at 85 percent of the moving previous five-year average of prices. The Senate conferees rejected the first two. The Senate Republican conferees accepted the third, but the administration representatives at the conference scuttled it by suggesting that the support level was probably too low.

In the end those senators who favored keeping permanent legislation for leverage in 2002 prevailed. Although some provisions of the 1938 and 1949 acts were repealed, many were merely suspended for the life of the 1996 Farm Act, including most of those applicable to price supports, the farmer-owned reserve, and emergency livestock feed assistance.[2] What was done in the conference regarding rice price supports illustrates the political power of rice producers. There was agreement to set the rice support price at $6.50 for the period of the act. However, the act also created a new permanent provision applicable to rice supports, but then suspended the provision for the life of the act. Under the new suspended permanent provision, the price support for rice is to be "not less than 50 percent, or more than 90 percent of the parity price for rice as the Secretary determines will not result in increasing stocks of rice to the Commodity Credit Corporation."[3] So one major commodity support price is still linked to the obsolete parity concept.

The price supports embodied in the 1996 Farm Act are still important because they represent a source of credit below commercial rates and a safety net against commodity prices dropping below the support levels. If commodity prices should drop below their support levels, marketing loans are triggered and producers are to receive payments based on the extent to which the prices are below the price support levels. These payments apply to total production, not just to historic yields on base acreages.

Dairy

The 1996 Farm Act also included tentative closure for U.S. dairy policy. Finally near the end of the House-Senate conference the conferees agreed to continue through 1999 federal government purchases of dairy products, but made no provision for purchases after the end of 1999, nor for any other method of directly supporting prices after that date. However, the USDA is to initiate a loan program for commercial processors of dairy products at the beginning of the year 2000. Although the interest rates charged will involve some subsidy to the processors relative to the rates they would pay if they borrowed commercially, the government is not to take any products as payment for the loans. In a sense the processors will have a line of credit that they can access continually to finance their inventories. Thus, so long as the law is not changed, starting in 2000 the federal government will not accumulate any government stocks as a way to support milk prices.

The practice of charging consumers more for milk that is sold for drinking than for milk that is sold for the making of cheese, butter, and nonfat dry milk will be continued into the future. However, the marketing order institutions designed to assure that this happens are modified. There were thirty-three of these institutions in 1996. The new act called for the number to be reduced to no more than fourteen, but not less than ten.

Also, the act includes consent for the then proposed northeast interstate dairy compact (until the marketing orders are reduced in number and reformed) with modest limitations on such things as how high the price of milk for drinking can be set—approximately $2.50 per gallon compared to the 1995 price of $1.88 per gallon.[4]

Peanuts

Unlike the programs for most of the commodities, the principal features of the peanut program were left intact. Thus, government regulation continues to be the norm for peanuts. The programs are governed with long-standing features including pools for quota and additional peanuts, area marketing associations, separate pools for Valencia peanuts produced in New Mexico, offsets within area quota pools, first and second uses of marketing assessments, farm poundage quotas, selling and leasing of farm poundage quotas, and limitations of transfers. There will continue to be two loan rates supporting prices—one for quota peanuts set at $6.10 per short ton, about 10 percent below the 1995 crop support rate, and one for nonquota peanuts to be set by the secretary of agriculture.[5] The complexity of this program now stands in stark contrast to the straightforward income transfers for most other commodities.

LEGISLATIVE LANGUAGE AND CONGRESSIONAL PASSAGE

With the official completion of the House-Senate conference on the 1996 Farm Bill, the immediate requirement was to adjust the legislative language to align with the perceived decisions of the conferees. Presumably this task is accomplished by the staff members of the committees and conference members. However, communications occur between the staffs and members and probably on occasion with lobbyists and department officials when implementation questions arise. In theory the requirement is to merely recall the conference decisions. However, reports indicate that some of the arguments among staff members over the details were every bit as intense as among the conferees, and in at least one instance even physically intense.

The focus of farm legislation activity then moved back to each chamber with an overwhelming expectation that each would approve it. They did.

The Senate vote on Thursday, March 28, was 74 yeas to 26 nays. Of 48 Democrats, 23 voted yea including Senator Leahy. Of the 52 Republicans, Arizona Senator John McCain voted no. Among the Democrats voting no were Dale Bumpers and David Pryor of Arkansas, Tom Harkin of Iowa, James Exon and J. Robert Kerrey of Nebraska, Kent Conrad and Byron L. Dorgan of North Dakota, John Glenn of Ohio, Thomas A. Daschle of South Dakota, and Russell D. Feingold and Herb Kohl of Wisconsin.

The House vote on the following day was 318 yeas to 89 nays. Of the 178 Democrats who voted, 106 voted yea including E. de la Garza and Charles W. Stenholm of Texas, and Gary A. Condit of California; of the 228 Republicans who voted, 17 voted no.

THE PRESIDENT SIGNS

The congressionally passed farm legislation was signed by President Clinton on April 4, 1996. However, he took the opportunity to express concern that the 1996 Farm Act that he had just signed did not provide for adjusting transfers from the federal government to producers (the Freedom to Farm flexibility payments) in response to price changes.[6] Secretary of Agriculture Dan Glickman expressed similar sentiments.[7] He made it clear that he would have liked to have seen increased attention to programs that responded to changes in commodity prices, that is, programs that were more generous to producers when prices were low and less generous when prices were high.[8] Neither President Clinton nor Secretary Glickman indicated how such programs should be funded. Nor did they criticize the prospective outlays of the new legislation.

PERSPECTIVES

With the signing of the 1996 Farm Act, the long, difficult task of crafting farm commodity legislation for 1996 and succeeding years came to a close. The policy of adjusting production from year to year in response to supply and demand conditions was abandoned as commodity prices continued to be bullish and optimism over export prospects continued.

In spite of expressed interest in balancing the federal budget, the long-term commitment to commodity programs involving transfers to producers and landowners was continued. This time the commitment was for seven years of fixed outlays for farm commodity programs. The total was recognized to be nearly equal to the total of the past seven years. While most observers expected that in 1996 and in 1997 these outlays would surpass what the outlays would have been if the 1990 Farm Act had been simply extended, it will be the later years that determine the overall welfare under the 1996 Farm Act.

Congressman Roberts, with shrewd and careful actions, had compromised just enough to attain approval of his Freedom to Farm concept of fixed payments to people who had received commodity program payments in the past. In the end, Congressmen Bob Livingston, R-La., and Joe Skeen, R-N.Mex., had been unsuccessful in sustaining their objections based on budget and jurisdictional considerations. In the Senate the need for sixty votes to close debate did not permit Senator Lugar to achieve his goals through the kind of brinkmanship that Congressman Roberts so ably employed in the House of Representatives. Senator Lugar forewent his proposal to explicitly phase out commodity programs. Instead he obtained the consent of the Senate for Congressman Roberts's proposal by accepting provisions critical to Senator Leahy. In the case of conservation, the provisions were important to Senator Lugar's wishes as well.

The role of the administration in a way paralleled how the president dealt with the budget conflict with Congress. It was also consistent with the way in which the congressional Agriculture Committees have tended to ignore Democratic and Republican administrations in the past when considering farm commodity legislation. At the same time some things might have been done differently and been more effective. It was not evident that the administration had given any one person authority to act decisively and quickly in negotiating with the Hill. This situation possibly reflected the conflicts within the administration over the appropriate policy to advocate. It also reflected the apparent undercutting of the secretary as Senate Agriculture Committee Democrats communicated directly with the White House rather than with the secretary at critical times during the debate. Finally, although the president's relaxed approach to budget matters had been successful in terms of keeping Congress at bay on budget reconciliation, the relaxed approach to farm bill negotiations seemed not to work as successfully. There was no political imperative to enact a budget reconciliation bill. In contrast, the incompatibility of the provisions of permanent farm legislation with its high price support rates ordained that there would have to be new farm commodity legislation.

The contrast between Senator Daschle's and Senator Leahy's beliefs and constituencies set up the most dramatic conflict of the entire farm bill chronicle. When Senator Leahy offered an alternative, Republican senators abandoned a compromise with Senator Daschle and his Senate colleagues. In spite of obvious earlier misgivings it was simply easier for the Republican senators to accept Senator Leahy's northeast interstate dairy compact, a title on conservation, and reauthorization of food programs than to compromise with Senator Daschle on flexibility payments and adjustments in farm commodity program benefits toward smaller producers and those with the greater financial needs. The Senate minority leader ultimately lost in his efforts with the White House and then in the Senate, even though these losses were not obvious until very close to the end of the Senate deliberations.

Through it all, the farm policy perception disconnect continued. The rhetoric emphasizing how farm programs are to help struggling operators who are financially strapped was used to justify programs that distribute benefits based on production volume that give the most to the largest producers of a selected group of farm crops and the largest farmland owners. Agricultural interests demonstrated their political power again to continue transfers from the federal government to farm producers and farmland owners.

A Future for Farm Commodity Legislation 12

We expect farm commodity legislation will be with us well beyond 2002. The economic health of the sector, at the particular time new legislation is considered, will have a major impact on both the design of new legislation and the dollars involved, just as it did in 1996. The condition of the federal budget will have less impact in spite of potential problems. Looking further ahead, the political imperative will continue to be to send money to agriculture and for the money to be distributed among producers and landowners based on production.

When the 1996 Farm Act was written, debated, and described, many people had the impression that farm programs terminated in 2002 along with the associated transfers to producers and farmland owners. But sunset provisions were never included in major drafts of the legislation. The closest any drafts came to calling for an end to commodity programs were sections that provided for elimination of permanent price authority included in the 1938 and 1949 farm legislation. In the final conference between the Senate and the House, even this proposed elimination of price authority was deleted. The 1996 Farm Act as passed suspends sections of the 1938 and the 1949 acts critical to the possible initiation of price supports. These 1938/1949 provisions would mean farm price support levels so high that no one expects that Congress would let them go into effect. Thus, the suspending amendment will force Congress and the administration to directly consider farm commodity legislation as the 1996 Farm Act expires, if not before.

What happens in 2002 will be a product of the perceived effects of the 1996 Farm Act as well as other forces including the federal budget, the economic health of agriculture, and the political power of various agricultural interests. We conclude that budget considerations will continue to be dominant in national politics, but that for commercial agricultural interests and for programs designed to transfer money to these interests, budget considerations will not be the determining factor. One reason is that commercial farm commodity pro-

125

grams are not large components of federal outlays. The other is that critics are not likely to have strong political influence over such outlays. Commodity programs are a case study of the potential leverage that can be exercised by a small group. The group is able to obtain benefits that are critically important to them by concentrating their efforts to gain what appears to others to be a very small and seemingly unimportant part of total government expenditure.

The economic health of the agricultural sector under Freedom to Farm will have an overwhelming influence on the design of new legislation. However, conditions in any specific year cannot be predicted, especially with the globalization of markets and the dependence of U.S. farm prices on growing demand around the world. Congress has tended to write farm bills in a reactive mode to deal with past and especially current problems, so the state of the farm economy in 2002 will have a critical effect on the direction of the next farm bill.

We expect the political power of the commercial agricultural interests—farm and nonfarm—to continue to be strong. Agriculture may even gain greater political strength as production is more concentrated and people with nonfarm interests have increased direct involvement with farm production. Environmental, rural development, and to some extent consumer interests will continue to press for modest appropriations. Commercial agricultural interests will accept increases in these areas grudgingly—possibly only when forced to do so. The USDA is not likely to become significantly more aggressive with rural development. It is having to become more so with respect to the environment. In the end we conclude that transfers to commercial farm producers and farmland owners will continue to be substantial beyond 2002.

Production flexibility contract payments like those called for in the 1996 Farm Act may be continued. However, another potential mechanism for substantial income transfer flows will be programs focused on risks associated with production and farm commodity prices. Such programs would permit the continued distribution of farm program transfers proportional to production of certain commodities. One question will be how these programs are crafted with respect to the distribution of benefits between producers who do not own the land they operate and the owners of the land. Such things as the terms of trade for different commodities and who is found to benefit from the new programs will influence the emphasis and coverage of alternative programs considered as modifications of or replacements for Freedom to Farm. Finally, the fact that Freedom to Farm is more transparent than previous agricultural programs may both encourage and change the character of the debate as we approach 2002. It is much clearer today who is receiving payments and how large these payments are.

THE BUDGET IMPERATIVE

The overall budget situation is still likely to be a driver of agricultural policy in 2002. One characteristic of the budget fix that Congress made in the 1990s is that a lot of the budget balancing pain was merely delayed. Neither Social Security nor Medicare/Medicaid were fully addressed. The budget surpluses of the late 1990s can only be used for tax cuts or other expenditures to the extent that dealing with the major entitlement funding shortfalls is delayed further.

A former staff member of the Agriculture Committee made the comment, after watching the process of the highway spending bill of 1998, that Congress has the ability to come up with funds for something it really wants. The implication was that more money could be found for agriculture in the future. However, in some ways the highway spending bill was almost unique. The prospect of a surplus in early 1998 allowed the dam of thwarted pork barrel spending to burst after being contained, to the frustration of Congress, for a number of years. The highway bill became a feeding frenzy where logrolling was possible on an unprecedented scale, and Congress seemingly sidestepped the requirement to find offsetting savings for excess expenditures in the highway bill. While there are opportunities for trade-offs to gain votes with agricultural legislation, they are nowhere near as extensive as they are with a massive public works bill that can reach into the heart of every congressional district. The question will be how successful the agricultural commodity program coalition is in shrewdly carving their programs out of a budget that again may be tight in 2002.

THE ECONOMIC HEALTH OF AGRICULTURE

The outlook for the agricultural economy was extremely optimistic when Freedom to Farm was passed. This optimism was based on rising farm commodity prices and fanned by those wanting to do away with land set-asides. They believed prices would continue to increase for agricultural commodities because of increased demand and in spite of increased land in production. In contrast, the long-range outlook by USDA showed declining real prices over the long term, but the optimistic viewpoints prevailed. The price declines occurring two years after the 1996 Farm Act challenged this optimistic view. While declining prices do not necessarily mean severe financial conditions in agriculture, they do heighten concerns and may lead to attempts to secure more of a safety net for the sector.

Over long periods of our history, agricultural prices have slowly declined in real terms (adjusted for inflation). The resource capacity of the agricultural sector, coupled with increased productivity associated with the discovery and adoption of new technology, has driven this increase in production relative to demand. Both the development and the adoption of technology were further stimulated when government programs supported prices above market-clearing levels. The beneficiaries of this increased productivity have been consumers on the one hand and the early adopters of new technology on the other. The slow real-price decline and the rewards for early adoption of new technology have driven producers towards larger consolidated farming operations to maintain or enhance their incomes. As these operations have prospered, their owners and operators have also become more politically astute and played a more active role in commodity legislation and implementation. This group's perceptions about how they are doing economically loom large in the debate over the efficacy of farm commodity programs.

The American agricultural sector is in overall better health financially in the late 1990s than it was in the mid-1980s at the beginning of the farm financial crisis. Balance sheets are in better shape for the majority of farms. Movement of people out of the agricultural sector to other jobs is easier; there are more second jobs available in the late 1990s than was the case in the 1980s after the Reagan recession when the Midwest was known as the rust bowl.

The concern of the administration and some Democratic legislators about variability of commodity prices and farm incomes is likely to resurface. Historically, price variability was seen as one of the main reasons for the government's involvement in agriculture. Secretary of Agriculture Henry Wallace used the biblical reference of Joseph, who planned for the cyclical variation of crops in Egypt (seven years of plenty followed by seven years of famine) to sell the concept of government storage under the "ever-normal granary." In the end, Sen. Thomas A. Daschle, D-S.Dak., and his colleagues in the Senate who favored legislation focused on stability were unsuccessful in attacking the Freedom to Farm concept, which not only changed the countercyclical role of payments but also ended land set-asides that were intended to reduce production when USDA believed stocks were too large. One reason for the failure of Daschle's approach was the optimism over commodity prices in late 1995 and early 1996 and the realization by producers that commodity payments in 1996 and 1997 would be less under the countercyclical approach than under Congressman Pat Roberts's, R-Kans., Freedom to Farm. If it is perceived that commodity producers are worse off in the early years of the twenty-first century under Freedom to Farm, some of the blame will fall on increased variability and the lack of countercyclical government programs—whether deserved or not.

THE POLITICAL POWER OF THE AGRICULTURAL COALITION

The political power of agricultural interests, both farm and nonfarm, will still be the key factor shaping what happens in 2002 or before. For Freedom to Farm this coalition included producers who believed they would be better off with fixed payments and fewer controls, agribusiness that wanted full production, and environmental/conservation groups that wanted to preserve and perhaps strengthen the environmental/conservation features from the 1985 and 1990 farm bills. If there is sentiment again among producers to embrace limitations on crop acres as a means of controlling supply to help increase low prices, the agribusiness interests could be very much at odds with producers in 2002. Environmental groups will have concerns about the efficacy of existing conservation programs to cope with mounting evidence of the negative impacts of agriculture on the landscape and in the waterways. If they continue the same course, environmental groups will press for program benefits to be linked to constraints or incentives that encourage farmers to lessen the impact of agriculture on the environment. As in the past, environmental groups will press for a continuation of agricultural programs with enough benefits to offset costs of compliance and to encourage farmers to join the program and meet the environmental and conservation standards.

The approach of environmentalists has been to attach their concerns to agricultural programs rather than to try to obtain their own appropriations and separate programs dealing with agriculture's environmental problems. Among other things, this means that the costs of environmental restrictions are not always obvious. In the 1985 Farm Act, certain environmental standards became conditions for receiving program benefits. The result has been relatively modest environmental programs, to some extent because commercial agricultural interests are not willing to make commodity program benefits conditional on meeting rigorous environmental standards.

The commercial agricultural coalition did hold together well in the period after the passage of Freedom to Farm. Attempts to reduce the production flexibility contract payments for producers and landowners were beaten back. The new agricultural leadership in the House appears to have grasped the reins left free with Congressman Roberts's move to the Senate. One interesting factor is the promise, if the agricultural sector is in trouble, to reopen the issue of the amount transferred to agriculture that was extracted from the House leadership in May 1995. This promise had a three-year time horizon. One question is whether the large additional payments to farmers under the omnibus budget bill of fall 1998 were part of that promise.

Impacts on the Setting for the 2002 Debate

Real agricultural prices are likely to be continuing a long-term downwards trend. The more relevant question will be whether commodity prices are declining or increasing in the short term as we near 2002.

The separation between ownership and operation of agricultural production assets will increase. Operators will own less of the productive assets, so returns to these assets will be less likely to be returns to farm operators. Yet, it is the operators who receive public empathy in times of poor economic conditions in agriculture. The question will be whether this increasing disconnect affects the debate and the level of public support for agriculture.

High tech inputs are likely to become more critical to the production process. Patenting of plant and animal genetic material and other innovations allows nonfarm agricultural interests greater profit potential. These input costs will have to be accepted by producers. Aggressive farm operators will be early adopters, pay for high tech inputs, and in so doing sell their skills, machinery services, and labor at modest-to-low levels of return. The new inputs coupled with information technology will allow agriculture to organize industrially with the potential for greatly changing control, profitability, and the distribution of returns.

The owners of agricultural assets, whether landlords or owner/operators, will have a vested interest in keeping a favored position in the stream of benefits from agricultural programs. It will be in their best interest to argue for programs that broadly capitalize program benefits into land and oppose programs that address operator income needs and disparities more directly.

Different commodities are likely to fare very differently under the more open market conditions of Freedom to Farm than they did under previous programs. A change in relative positions does not go unnoticed, and usually results in political action. Wheat may do less well under Freedom to Farm because previous farm programs redressed its less competitive world position. Rice and cotton may be in similar positions. Dairy lost a large degree of economic support under Freedom to Farm that had been part of previous programs. The question will be whether free market opportunities and the northeast dairy compact make up for this loss.

The desire to socialize risks and privatize gains will remain strong. This tradition from the American past was not scrapped with the post-1994 free market era in American politics. This approach is strongly supported by entrepreneurs and results in strong lobby pressure on Congress. As one example, the financial industry attempts to do this with changes in banking and financial regulations. Agriculture has done it through price supports, nonrecourse loans, and crop insurance, and presses on to do it with revenue insurance. There may be more of a reason to do this for agriculture relative to other industries given the strategic necessity and weather-related nature of the industry.

Have We Redefined the Role of Government in Agriculture?

The 1996 Farm Act may reflect a change in the basic view of the public about the role of government in agriculture. The congressional election of 1994 swept in a majority that was determined to reduce government's role across the board, and agriculture was one target of this effort. Yet, we passed a farm bill transferring income to agriculture with fewer obligations on the part of the producers and farmland owners than before while cutting nonfarm welfare payments to other groups at the same time. The new farm bill did greatly reduce government's role in the production process, but neither Congress nor the public have demonstrated that a radical change in the willingness of the general public to make transfers to producers and farmland owners has taken place in spite of a different approach to the income transfer process—one that no longer requires government micromanagement of production. The basic transfers have been continued without a strong public case being made for the old rationale or for a new rationale for such transfers.

Is there a new case to be made for income transfers to producers and farmland owners? Are there parts of the old rationale that still hold today and make good public policy sense? Sen. Richard G. Lugar, R-Ind., as chairman of the Senate Agriculture Committee, attempted to initiate a debate along these lines in 1994. His call for phasing out commodity programs dramatized his concern about keeping programs that may have lost some of their justification. The ensuing legislative activity, however, ignored the Senator Lugar-type questions as to whether federal farm commodity programs are justifiable given federal budget conditions; how farm resources are organized, managed, and controlled today; and the role of U.S. agriculture domestically and in the world economy. The new industrial organization of agriculture also raises questions about the extent and effect of these programs on efficiency as well as equity. Is the public better served with an industrialized agriculture, do agricultural programs speed or impede our movement in that direction, and are we comfortable with programs where the income transfers appear to be directed more to the better-off producers?

We must wait and see whether these basic questions about agricultural programs and the role of government in agriculture that Senator Lugar tried to initiate in 1994 will be asked and finally discussed as Congress decides what to do in 2002.

A Primer on U.S. Agricultural Policy Before 1996

13

Prior to the 1920s and 1930s the involvement of government in agriculture was indirect. From the outset, the primary concern was land policy and the settling of the continent to extend political and economic control. Over time this included the encouragement of transportation, from canals to the transcontinental railroad; the encouragement of new technology such as widespread seed distribution by the Department of Agriculture after the Civil War; and the encouragement of practical education in the industrial and agricultural arts with the Morrill Act and the founding of the land grant universities. Farm groups, such as the Grange, gained political power representing farm interests and successfully promoted antitrust and regulatory activities to curb the excesses of railroads and other powerful interests. Farm groups also became concerned with the disparity in income between farm and urban people. This became a major justification for farm programs as these groups gained political power and public support for assistance to agriculture.

One hallmark throughout has been the vast production potential of the country. Production has almost always outstripped demand to the benefit of the consumer. Another hallmark has been our relative abundance of land. Labor was the scarce resource, and for most of our agricultural history, the technology induced by this scarcity was labor-saving technology. Thus, it was the cotton gin, the mechanical reaper, and the tractor that marked U.S. agricultural technological progress until the early 1900s. We were less marked by the intensification of cropping that occurred in land-poor Europe and Japan.

For a brief period in the early 1900s and during the First World War, there was a golden age for American agriculture. The frontier had closed, agriculture was no longer expanding output through the cultivation of new lands, and the expansion of immigration to the industrial cities of America added demand for food that was further intensified by the First World War and the demands from the Allies in Europe. The cry then was that "wheat will win the war," and production expanded dramatically. The concept of parity for agriculture is rooted in this time period when the value of agricultural products expressed in terms of what the farmer needed to buy reached a high that has not been achieved again. From this comes the index of parity prices that is part of the 1949 Farm Act and many previous ones.

The agricultural economy collapsed in 1920–21. Demand for U.S. agricultural products from Europe ground to a halt as production there resumed its normal course. This is when the depression began for agriculture. Agricultural prices collapsed around the world. The U.S. government entered the market directly in the late 1920s and tried schemes to purchase commodities to raise their prices, most notably cotton. These ended in failure as farmers grew even more to sell at the higher price. The financial collapse of 1929 further eroded the economic position of farmers. A major argument at the time was whether the poor position of farmers was caused by the bad macroeconomic policy of the United States that contracted the money supply, the demand for goods, and capital for investment, or whether it was caused by overproduction. In either case, the early attempts to support prices by government purchases were less than successful. They solved neither the macroeconomic problem nor the oversupply problem.

During the Roosevelt administration, agricultural assistance programs were developed that have continued in one form or another since then. The Agricultural Adjustment Act of 1933 set prices and tried to control supply. The Supreme Court struck down provisions of this act, and new acts in 1937 and 1938 set the mechanisms that functioned to control supply and support prices for roughly the next sixty years.

A key feature of the program was the nonrecourse loan. Henry Wallace, secretary of agriculture at the time, described this in terms of the biblical story of Joseph and his management of grain stocks for the seven years of plenty and the seven years of famine. The government set a loan rate for a commodity that was covered by the program. At harvest time the farmer could take a nonrecourse loan on his crop based on this loan rate value. If the market price was above the loan rate when the loan came due, the farmer could pay back the loan and pocket the extra from the higher price. If the market price was below the loan rate, the farmer could walk away from the loan and the government took ownership of the crop that had been pledged as security for the loan. In this way the government built stocks (an "ever-normal granary") in years of low prices and could sell stocks in periods when prices were higher than the loan rate. This was first seen as a way to even out seasonal price fluctuations, but gradually developed into a multiyear stabilization effort that involved the active accumulation of government-owned stocks. This system placed a high value on price stability for agriculture and for consumers.

Not all commodities were covered by government programs. The major commodities of corn, wheat, cotton, peanuts, rice, and tobacco, along with sugar, some minor grains, and even wool and honey, were included. Not all commodities have been or are treated equally in terms of the relative levels of government support. After an initial attempt to raise hog prices, livestock was not part of government programs, but milk was. Some specialty crops and fruits were allowed to enhance prices under marketing orders.

Key to the commodity programs was the level of the loan rate. If the loan rate was set higher than the average long-run price, the government would tend to accumulate stocks. Concern with the welfare of producers, whose income at the depths of the depression was 40 percent of the income of the urban population, led Congress to transfer income through the nonrecourse loan program by having a high loan rate. Stocks tended to accumulate and programs became expensive.

There was a concurrent effort to control supply. Initially this was done on a mandatory basis until struck down by the courts. Then programs were made voluntary in the late 1930s and operated under similar structure until 1996. The price protection of the nonrecourse loan and other benefits were offered to farmers if they joined the program, which included an obligation to control supply, usually through limiting the land in production. It is here that discretion was given to the secretary of agriculture to declare set-asides or ARPs. If supply was determined by USDA to be more than adequate for the coming crop year, the secretary would require those who joined the program not to plant a portion of their acreage. In addition to the set-aside acres, land conservation programs such as the lands returned to grass after the dust bowl in the 1930s; the Eisenhower soil bank, which enrolled up to twenty-nine million acres in the 1950s; and the CRP, which enrolled up to thirty-six million acres in the 1980s and 1990s, were also efforts to control production as well as to protect and improve the land that was fragile and in some cases should not have been farmed. The early conservation programs of the 1930s were also seen as a way to help boost cash income to farmers with payments to idle conserving acres and payments for farm conservation improvements. The dilemma remained whether these conservation programs that set land aside for conserving uses and paid the farmer were income transfers, supply reduction measures, or conservation measures. There are no land set-asides or ARPs to limit supply in Freedom to Farm. The CRP that sets land aside for conservation and environmental quality continues with a renewed focus on conservation.

During the Second World War, the war economy demanded all that agriculture could produce. Producers were concerned that there would be a repeat of the postwar collapse of 1920–21. Thus, support prices were set relatively high in the post-World War II period. In the 1950s, the mechanism of the nonrecourse loans and other price support mechanisms resulted in large accumulations of grain stocks (and at times butter, cheese, and a variety of other commodities). In the 1970s a target price mechanism was put in place, and the farmer was paid directly the difference between the market price (or the loan rate if it was higher than the market price) and the target price. On this basis the government wrote a direct price support payment check to the farmer—something that would have looked inappropriately socialist in earlier times. This was a simpler and probably less expensive way of transferring money to farmers. Under the alternative where the government paid a high loan rate to farmers, the government was put in the position of accumulating costly stocks, supporting world prices, and encouraging farmers in other parts of the world to produce more than they would have otherwise in competition with American farmers. With a target price and a much lower loan rate U.S. farmers received income support without the United States supporting world prices.

The traditional U.S. commodity program was countercyclical by design. The notion was that when prices were very high producers would not put their crops under loan and would be unlikely to receive a target payment. Expenditure for programs would be very small. However, in years of bumper crops, low demand, and very low prices, commodity programs could be extremely expensive as they were in 1986/87 at the time of the farm financial crisis. Expenditures could be a few billion dollars or in the tens of billions of dollars. This gave heartburn to the budget planners. In essence, the traditional commodity programs were open-ended entitlements, like Social Security, where payments had to be made to meet the contractual terms of the program. The budget-makers never knew quite what agricultural programs would cost. Changes in supply due to weather were the major swing factor. Domestic demand was relatively stable. Export demand became a bigger factor in the 1970s and 1980s as it had been at the time of the First World War. In retrospect, the use of land set-asides to "manage" production was far from perfect because the secretary of agriculture couldn't predict weather or future exports perfectly either. The Freedom to Farm program is not countercyclical. There are no production controls to hold surpluses down, and the payments do not vary inversely with price.

The incidence of cost of U.S. agricultural programs is largely on the taxpayer. While U.S. government programs do enhance prices, most of the extra income to producers comes through direct payments. This is in contrast to agricultural support programs in much of Europe and Japan, where price enhancement forcing consumers to pay more for food is a large part of the income transfer that goes to producers as a result of public policy. There is an important equity issue here. In the United States the lower prices for food are favorable to low-income citizens and, if income taxes are progressive, higher-income citizens pay more of the cost of maintaining the production system that provides low-cost food.

Finally, a note on process. The work in Congress by Agriculture Committee members and by those members interested in agriculture tended to be very bipartisan in the past. Admittedly there have been disagreements along party lines, some of them bitter and personal. Republicans were very critical of many New Deal farm programs in the 1930s. Democrats rejected the Brannan plan of 1948, which targeted transfers more to producers, and opposed the free markets advocated by Secretary of Agriculture Ezra Taft Benson under Eisenhower in the 1950s. However, agricultural policy became increasingly bipartisan as members of Congress with an interest in agriculture realized they were dealing with a generic problem of U.S. agriculture's overcapacity leading to politically unacceptable low prices. Consequently, bipartisan support evolved for transfers to the agricultural sector including the price and income supports, measures to restrain production, and measures to enhance demand like export subsidies, food stamps, school lunches, and even ethanol production. Other members of Congress not concerned with agriculture were willing to let the aggies do it because agricultural production was a small part of the economy and of the business of Congress. However, the year 1995 was dramatically different. The ironclad budget split Republicans from Democrats, commodity groups that previously cooperated to get more for all players fought for their piece of the fixed pie, and members of Congress outside the agricultural circle became more interested in cutting agriculture and food programs to gain budget for their own interests. The 1996 Farm Act was made in a very different dynamic from previous ones.

Some of the nomenclature and details that are often referred to and important to a basic understanding of agricultural programs include

—Base acres: A farm participating in traditional farm programs established a given number of acres historically devoted to a specific program commodity. Under the program farmers could not plant more than the base acres in that commodity (an essential for supply control). Base acres also enhanced the value of the land—representing a stream of price-support payments that became capitalized into the value of the land.

- Flex acres: This provision from the last several traditional farm programs gave farmers the opportunity to plant another crop of their choice on a portion of their base acres. Expansion of this provision was one alternative suggested for the 1996 Farm Act. Freedom to Farm gives freedom to plant most other crops. Base and flex acres become meaningless.
- Payment limitations: The commodity programs were geared to volume of production, not to a farmer's income. Producers received price support from the loan rate or target price-support payments calculated by multiplying their base acres times their normal yield times the loan rate or price support. A very large farm would receive a very large payment. At various times since the 1970s limitations were put on the total value of payments that a producer could receive. As farms were restructured to avoid this restriction a triple-entity rule was added to limit the ability to collect from many different units that might actually be commonly owned.
- Mandatory crop insurance and disaster payments: Congress never saw a disaster whose victims it didn't want to help. The instances of flood, drought, etc., usually result in victims receiving generous assistance from Congress. This is in spite of the existence of federally subsidized crop insurance. Congress has at times required crop insurance as part of the commodity program. While it would result in a broader and sounder actuarial base for the program, it has not been popular with farmers—especially in areas where there is less weather risk. Congress has also stated that it would not give specific disaster assistance in order to encourage participation and improve the coverage and actuarial soundness of the crop insurance program. However, farmers believe that Congress will respond to their pleas for help when disaster strikes. Congress has seldom proven them wrong and was responsive again in the fall of 1998.
- Conservation compliance: In the 1980s environmental groups pressed Congress to improve the performance of agriculture with respect to its impact on water quality, land degradation, and other key concerns. The conservation compliance standards were then included in commodity programs. In order for a farmer to receive program support, the farmer had to meet certain modest conservation standards. The rule was maintained with the 1996 Farm Act. Key here is that the commodity program has to be attractive enough for farmers to be willing to take on the extra responsibility of conservation compliance. The more that is required under compliance, the more attractive the program needs to be to get broad voluntary participation.
- Farmer-owned reserve: Under the traditional commodity programs

where the government would take title to grain under the nonrecourse loan, it became extremely expensive for the government to store this grain. The farmer-owned reserve was set up under which the government paid the farmer to store the grain on the farm. This was subject to rules about when the farmer could sell the grain, which was placed under nonrecourse loan for an extended three- to five-year period. Farmers were paid a storage fee and given low-interest loans to build storage facilities, which made it attractive for farmers to participate. This is eliminated under the 1996 act.

—Marketing loan: The marketing loan allows farmers to repay their commodity loans at rates below the loan rate when world prices are below the loan rate. This ensures that commodities can move on to world markets and not become government-owned stocks when world prices are low. The Freedom to Farm program includes marketing loans with loan rates set at low levels.

—The 1949 Farm Act: The 1949 act has been the mother of all farm acts since the 1933 Agricultural Adjustment Act until Freedom to Farm radically changed programs in 1996. It was passed right after the Second World War when farmers were afraid that agriculture would suffer another depression from a drop in demand just as it had after the First World War. The price supports in this act are thus very high and tied to the concept of parity purchasing power for farmers. A reversion to the 1949 act in 2002 was added to Freedom to Farm at the very end of the making of the 1996 Farm Act.

A longer-term perspective on where we are and where we have come from in agricultural policy in the United States is provided by Chester Davis and Howard Tolley, who served under Henry Wallace in Roosevelt's New Deal.

Davis argued that "a nation's agricultural policy is not set forth in a single law, or even in a system of laws dealing directly with current farm problems. It is expressed in a complexity of laws and attitudes which in the importance of their influence on agriculture, shade off from direct measures like the Agricultural Adjustment Act through the almost infinite fields of taxation, tariffs, international trade, and labor, credit, and banking policy."[1] While our focus has been specific agricultural policy, the nonagricultural factors Davis identified in 1940 have even more influence on agriculture than they did in 1940. Federal macroeconomic policy brought agriculture to its knees in the 1980s farm financial crisis, and nonagricultural policies can make or break producers and farmland owners in the future.

Howard Tolley, reviewing the progress made under the New Deal, noted that the objectives of farm policy were of three types: "(1) Activities designed to increase incomes of farmers who produce commodities for sale on a commercial scale; (2) the efforts to raise incomes and to improve the living conditions of migrant laborers, sharecroppers, subsistence farmers, victims of drought or flood, and others at a disadvantage within agriculture itself; and (3) activities designed to encourage better land use [including conservation] and more efficient production."[2] He then says that most government programs have been aimed at the improvement of commercial agriculture and that more government attention will be aimed at the last two. Our 1990s take on this 1940 assessment is similar to Tolley's. Our programs have been extremely successful in promoting commercial agriculture, but if the public is concerned about those at a disadvantage within agriculture and with the resource and environmental issues associated with agriculture, much remains to be done.[3]

Appendix 1
SENATOR LUGAR'S QUESTIONS ON PROSPECTIVE FARM POLICY

DRAFT QUESTIONS FOR COMPREHENSIVE SENATE AGRICULTURE COMMITTEE HEARINGS ON THE 1995 FARM BILL

SENATOR RICHARD G. LUGAR
CHAIRMAN

PREPARED BY THE REPUBLICAN STAFF OF THE SENATE COMMITTEE ON AGRICULTURE, NUTRITION AND FORESTRY

DECEMBER 9, 1994

COMMODITY PROGRAMS

1. How would one allocate Federal agriculture outlays among program areas such as market development, crop assistance, food assistance, etc., if the criterion were to maximize public return on investment and enhance the competitiveness of U.S. agriculture in the world market?

2. Which activities of the following agencies could be streamlined, privatized or satisfied better by others:

    Agriculture Marketing Service, Animal and Plant Inspection Service, Agricultural Stabilization and Conservation Service, Extension Service, Foreign Agriculture Service, Federal Crop Insurance Corporation, Farmers Home Administration, Food and Nutrition Service, Forest Service, Federal Grain Inspection, Packers and Stockyards Administration, Food Safety and Inspection Service, Rural Electrification Administration, Food and Drug Administration. (NOTE: Agency names prior to reorganization are used because of their familiarity.)

3. One rationale for farm programs is a reduction in price volatility. Are there adequate opportunities through the private sector through futures and options to provide alternative risk management approaches for farmers? Do farm programs substitute for these risk management strategies? Since beef, pork and poultry prices at the retail level are not

extremely volatile despite the absence of programs to support these commodities, are programs really necessary to reduce retail price volatility?

4. Why are Acreage Reduction Programs sound public or fiscal policy when they require farmers to idle productive land and spread their fixed costs over fewer acres, negatively affecting their ability to turn a profit?

5. If a goal of farm policy is to keep family farmers on the land, has the current program helped or hindered this goal? How has the rate of exit from agriculture compared to that in both more and less subsidized farm sectors in other countries? What is the failure rate for farmers who participate in the farm programs versus farmers who grow unsubsidized crops? Why should taxpayers subsidize farmers when they do not subsidize small businesses, which have a failure rate hovering around 50 percent?

6. What level of gross income would the average producer of program crops need in order to have full time employment from that operation? What percent of our total farm production comes from farms of this size or larger? How dependent on off-farm income are producers below this level?

7. With respect to both the dollar value and number of recipients, do price support benefits go primarily to full time or part time farmers?

8. Would our nation run a serious risk of losing its abundant food supply if commodity programs did not exist?

9. What impact would elimination of farm programs have on land values? Given current debt to asset ratios for agriculture would this have a substantial impact upon farmers' ability to obtain substantial financing?

10. Many people believe that current farm programs encourage farmers to bring marginal land into production or maintain it in cropping. Without farm programs, would farmers be more or less likely to do this?

11. What is the rationale for subsidizing some crops but not others? Is there evidence that producers of non-subsidized crops have prospered less than producers of subsidized crops? What is the rationale for transferring public funds from taxpayers, most of whom have moderate incomes, to all farmers, including those whose incomes and wealth are substantially above the national average for all Americans?

12. Why is subsidized crop insurance and disaster relief appropriate for agriculture, and not for other sectors of the economy?

13. What impact would elimination of the tobacco program have on the price of tobacco products? What impact would this then have on smoking? Why should there be any government involvement in the production of tobacco?

14. Since passage of the marketing loan program for cotton and rice in 1985, has the cost of the programs been higher than would have been the case if the traditional loan rate program had continued? By how much? Some economists project that because of the marketing loan program, a rise in world cotton prices leads to a decline in U.S. cotton farmers' income. Is this true, and if so why?

15. Why should sugar production be protected and imports restricted if the result is higher sugar prices for American consumers?

16. Since peanut butter is an inexpensive source of protein for many lower and middle income Americans, does it make sense to limit the amount of peanuts that can be sold within the U.S., preventing consumers from purchasing peanut butter at lower price? Do these limitations on marketing peanuts for domestic edible use have a negative effect on farmers who do not own or lease production quotas?

17. Many agricultural economists say milk marketing orders result in price discrimination, that milk for fluid uses receives a higher price than milk used for manufacturing. Does this make sense? Should there be a difference in price for different uses of milk? How does this practice affect prices for consumers?

18. With advances in technology and transportation, why is a system of milk marketing orders necessary to ensure adequate supplies of fluid milk throughout the country?

19. What are the most prominent examples of agricultural programs that may confer subsidies to commodities other than those for which they are directly designed? For example, do corn farmers derive a benefit from the operation of the sugar price support program? Does the associated production of high fructose corn sweetener have a negative effect on the demand for soybean meal?

CONSERVATION PROGRAMS

1. Is there a need for the swampbuster provision given the existence of Section 404 of the Clean Water Act?

2. Should a CRP be continued? If so, at what size cost? Should the criteria for entry be expanded or modified to include water quality, wildlife habitat or other items? If so, how should the criteria be structured?

3. If the receipt of farm program benefits were not tied to conservation practices, would there be a significant decline in the amount of acreage on which conservation measures occur?

4. What is the most cost-effective strategy for ensuring the environmental sustainability of U.S. agricultural production capacity?

EXPORT PROGRAMS

1. Under the Export Enhancement Program, selected foreign customers pay less for U.S. grain than other customers, including U.S. domestic users. Who benefits and who loses from this policy? What is the justification for transferring taxpayer funds to some foreign buyers but not others, and not at all to domestic buyers?

2. In the early years of the EEP, subsidies were paid in government-owned commodities, a practice that increased total supply in the marketplace, presumably affecting total consumption. More recently, as government

stockpiles were exhausted, subsidies were paid in cash. In such circumstances, has EEP had any effect on total consumption?

3. Does EEP reduce world wheat prices below the level they would otherwise attain? If EEP reduces world wheat prices, is it logical to expect that it has either reduced U.S. corn exports or reduced corn prices to maintain market share? Similarly, if wheat prices have been reduced, would Canadian and South American producers have an incentive to switch plantings to oilseed crops (canola and soybeans), thereby increasing total supplies and reducing prices of these crops for U.S. farmers?

4. Since the beginning of the EEP in 1985, what have the U.S. and EU share of world wheat markets been? Is there evidence that the EEP has either increased U.S. market share or prevented a further erosion of that share? If the answer is that the EEP was intended to induce the EU to bargain, is there any justification for keeping the program now that the Uruguay Round has mandated a 21% reduction in subsidized EU export volumes?

5. Does U.S. wheat market share since 1985 correlate more closely with EEP sales, or with short crops in competing nations, including Canada and Australia?

6. Has the EEP bonus and price review process within USDA, as well as USDA's role in the allocation of EEP eligibility to various countries, substantially increased the federal government's role in wheat markets? Specifically, are market price movements more closely affected by government action under the EEP than they were before its creation, and does the private sector devote more resources to monitoring and attempting to anticipate government actions than before the creation of EEP? What are the costs to the private sector of this development?

7. Under a Uruguay Round regime in which total subsidized exports cannot rise, but must fall, is it likely that the EEP can increase U.S. market share? Is it more likely that the primary effect of subsidies will be to lower world prices?

8. As permissible subsidized volumes sharing in coming years, how will USDA allocate the smaller subsidy pie among the large universe of

claimants? Will the Department reduce eligible purchases by all customers equally, completely exclude some customers while allowing the purchases of others to remain the same, or pursue some other course? Is there any evidence that USDA has considered its best course of action, presented options to policymakers, or analyzed the likely outcome of the coming rationing of subsidies? If so what conclusions has USDA drawn?

9. As long as U.S. loan rates are below market-clearing levels, what would prevent U.S. prices from falling to whatever level was necessary to compete with the EU, even if the EEP were abolished?

10. Why has the most rapid growth in U.S. food and farm exports occurred in products that receive few if any subsidies—meat, poultry, pet food, snack foods, fruits and vegetables, wine? What does this say about the effectiveness of U.S. export subsidies?

11. What evidence is there that the GSM-102 program has expanded total import demand? Is the program internally inconsistent, featuring both an implicit subsidy (the extension of credits to customers who would not receive the credits in the absence of federal guarantees), and simultaneously a statutory requirement that all buyers be creditworthy?

12. What are the merits of the General Accounting Office's argument that the P.L. 480 Title I program has been ineffective in promoting broad-based sustainable development, largely because it is too small to make much difference?

13. What progress has been made in evaluating the effects of market development and market promotion? Are there ways to evaluate the effectiveness of these programs besides simply looking at before-and-after export levels, which could have been influenced by many other factors?

NUTRITION

1. Which of USDA's 16 nutrition programs could be consolidated? Which are most suited to a block grant approach, and why? What would be the impact of block granting or consolidation on participation, administrative costs, and program effectiveness?

2. What capability does the Department of Agriculture have to assess and quantify the extent of benefit diversion and fraud in the Food Stamp Program? What data have been yielded in this area by the Electronic Benefit Transfer system in place in Maryland?

3. What would be the program impacts of cashing out food stamp benefits for only those households participating in welfare reform-related employment and training programs?

4. What is the best way to make employment and training under the Food Stamp Program more effective? Should it be consolidated with other employment and training programs, for example, or should its rules be brought in closer conformity with AFDC work provisions?

5. Would Electronic Benefit Transfer be a cost-effective delivery system if the Food Stamp Program law and regulations get in the way of the states' welfare reform initiatives?

6. What specific provisions of Food Stamp law and regulations get in the way of states' welfare reform initiatives?

7. Given that, because of limited funding, not everyone who is eligible for WIC can participate, what is the rationale for not requiring applicants to provide documentation of their income?

RURAL DEVELOPMENT/FMHA PROGRAMS

1. Is the Farmers Home Administration needed to encourage replacement of retiring farmers? Why? What level of outlays are needed? If needed, can commercial servicing of loans and guarantees reduce transaction costs and improve customer service?

2. One goal of some USDA programs is to broaden the economic base of rural America beyond production agriculture. To what extent have these rural development efforts been successful? Are there more efficient means of encouraging the private sector or non-federal government to achieve these same goals?

3. According to the Congressional Research Service (CRS), the Rural

Electrification Administration (REA) originally was designed to provide unsubsidized credit for rural electrification—subsidies crept in later. Given the high rate of rural electrification, is there a need for subsidies for all or some REA borrowers?

4. REA supporters argue that rival suppliers of electricity also receive government support, albeit less directly. Moreover, they argue that efforts by rural electric cooperatives to attract businesses and jobs to its service area often result in annexation by municipal power systems, thus cherry-picking the most lucrative parts of a coop's service area. Do the coops have a point and, if so, how should the problem be resolved? Is the support for none-REA suppliers of electricity comparable to that given to REA borrowers? Should the federal government provide any support, direct or indirect, to any supplier of electricity?

5. Critics of USDA's rural telephone program complain that once a rural telephone company has qualified to borrow from the government, it remains eligible even if it has been purchased by a large commercial company. What is the justification for this situation?

6. The Farmers Home Administration has been criticized for having written off billions of dollars of farm loans. FmHA's farmer programs have been plagued by delinquencies. What has caused this problem? Are there ways to restructure FmHA's programs to limit losses to taxpayers?

7. For borrowers in financial distress, FmHA tries to ensure that borrowers can remain in farming. Do historic rates of return on investment in agriculture justify a policy of maintaining even indebted farms in operation?

8. Does it make sense to make new loans to borrowers after having reduced or forgiven their previous FmHA debt? Is the FmHA borrowers' appeals process too cumbersome and can it be streamlined to operate more efficiently for borrowers and taxpayers?

9. Is adequate credit available in rural America not only for production agriculture, but also for rural business development and infrastructure

improvement? How do the interest margins of both banks and the Farm Credit System today compare to the past? Interest margins are one component of earnings. What happens to these earnings? Do they remain in the rural sector? Are interest rates in rural areas higher than those in urban areas? If so, why?

10. The Federal Agricultural Mortgage Corporation, or Farmer Mac, was created in 1988 to develop a secondary market for agricultural loans in order to attract new capital for long-term agricultural financing and to provide greater liquidity to agricultural and rural lenders. However, its volume of activity has been much lower than expected. Is an independent secondary market important to agriculture or rural America? If so should Farmer Mac's statutory authority be changed to increase its efficiency and business volume or should an alternative or duplicate secondary market be created? If Farmer Mac continues, should it be used to carry out any other agricultural credit or rural development activities?

Appendix 2
Early 1995 Anonymous Draft Approach to Withdrawal of Government Farm Programs

Possible approach to major withdrawal of Federal U.S. Government farm and farm product regulations and programs designed to restrain production, restrict market supplies and increase prices.

Transition Period

Ten years 1996-2005. Adjustments in programs so that at the end of the ten year period the Federal U.S. Government is no longer substantially involved in restraining production, restricting market supplies, and increasing prices of farm and farm related products. It is envisioned that by 2005 no Federal programs would restrict farm production on an annual basis. Quantities of farm products moving through marketing channels would not be restricted. In 2005 the direct government support of farm prices would be no more than fifty percent of 1995 support prices. Government recourse loans would be available at unsubsidized interest rates and consistent with support price levels. Non recourse and marketing loans would not be available. The Government would own and control stocks of selected commodities in amounts decided by the Congress.

More specific steps that are suggested to achieve the withdrawal by 2005 include:
Commodities now regulated with price supports, ARP's, base acreages and base yields. (Wheat, feed grains, cotton, rice, and oilseeds)

> Support Prices and Commodity Loans. Reduce support prices each year by 5% of 1995 support prices until the year 2005 at which time the support prices will be 50% of the 1995 support prices. Make recourse loans (at the then current support prices) available in the year 1996 and succeeding years. Eliminate the non recourse loans.

> Deficiency Payments. For each ASCS farm establish 1992, 1993, and 1994 annual average deficiency payments for all crops (1993-94 averge [sic] deficiency payment). "Attach" ½ of the 1992-94 average deficiency payment base to the "farm" and permit it to be transferred with the land if sold. "Attach" the other ½ of the 1992-94 average deficiency payment base to individual operator of land in 1992-94.

Starting in 1996 make deficiency payments to the respective holders of the 1992-94 average deficiency payments bases but at amounts successively reduced by 10% each year. For example, in 1996 the amounts paid by the USDA will equal 90% of the base amounts, in 1997, 80% until in 2005 they are 10% of the base amounts and in 2006 they are zero.

Target Prices. Eliminate target prices since there will be no direct payments other than those based on deficiency payments in the three years 1992-94.

Government Stocks. Discontinue the Farmer Owned Reserve by 1998. Congress decides the level of stocks to be held in a government stocks program. Purchase stocks in the open market. Sell one-third of the stocks each year and replace with purchases unless Secretary of Agriculture with approval of Senate and House recommends that they not be replaced.

Other: Producers will be free to plant any combination of crops. On individual farms the acreages of individual crops will not be limited. Payments of deficiency payments and availability of recourse loans will not be affected by which crops are grown. Eligibility to receive deficiency payments will continue to require compliance with the conservation plan for the respective farm.

Commodities now regulated with a combination of price supports and marketing agreements (dairy).

Reduce milk support price each year by 7% of 1995 support price until the year 2005 at which time the support price will be 50% of the 1995 support price.

Eliminate by 1997 all restrictions except sanitary and health restrictions on the movement among states of all dairy products including milk and milk products for fluid milk consumption.

Phase out all marketing orders by the year 2000.

Commodities now regulated with a combination of price supports, limitations on individual farm production, and import quota (sugar, peanuts, tobacco).

Reduce support prices each year by 7% of 1995 support price until the year 1998 at which time the support price will be 79% of the 1995 support price. In preparation for announcing support price for 1999, the USDA shall make a forecast of the free market price levels in the year 2006. On the basis of this estimated price for 2006 reduce support prices by an equal amount in each year 1999 through 2006 so that in 2006 the support price is no more than 70% of the forecasted free market price.

Increase import quota by 10% in each year.

Relax restrictions on production of individual farmers by increasing any and all quotas and acreage restraints by 10% each year.

Conservation Reserve Program

Permit all CRP contracts to expire.

As contracts expire place 50% of annual CRP payments associated with the expiring contracts into block grants to be transferred to states and or watershed authorities. The block grants to be used by state and watershed authorities in ways that achieves [sic] environmental improvements including reduced soil erosion and reduced pollution of surface and ground water and meet the approval of the Soil Conservation Service. Groups that are vested with authority on how to use the block grant money must be elected officials. At least some of the members must be current farm producers. No member of the decision making groups or their direct relatives may receive any of the funds.

Ten Year Transition Period From the Present Commodity Price Support, Marketing Loans, and Deficiency Payments

All who have received checks (deficiency payments) over the past three years are automatically on the list to receive checks during the transition period.

Similarly, all who have received non recourse commodity loans over the past three years are automatically on the a list to receive non recourse loans during the transition period.

For each recipient of checks over the past three years the three annual amounts will be averaged and the resulting number will be their base check amount.

No entity or person who did not receive a check or a loan in the past three years will be eligible to receive such during the transition period.

The three year base amounts will be partially attached to the land and transferable and partially attached to the recipient and not be transferable.

During the first year of transition the 90% of the base amounts will be attached to the land and transferable; 10% will be attached to the recipient and not be transferable. In the second year the percentages will be 80% and 20%. In the third year the percentages will be 70% and 30% etc.

Price supports levels will be reduced 10 percentage points each year during the transition period.

Checks as a percentage of the base amounts will be decreased 10 percentage points each year. As will the maximum of nonrecourse loan amounts.

Appendix 3
The Summer 1995 Anonymous Draft of "The Freedom to Farm Act of 1995"

THE FREEDOM TO FARM ACT OF 1995

A Seven Year Contract with Production Agriculture

A Brief Description

Farmers who have participated in government programs in three of the last five years would be eligible to enter into a seven year contract with the Federal government to receive an annual payment based on a percentage of their historical payments (based on payments received from 1990 through 1995) from the Commodity Credit Corporation. In exchange for the payment, farmers would be required to maintain previously developed NRCS conservation compliance plans. Acreage reduction programs (ARPs), annual set-asides, and paid diversions would be eliminated. Farm and crop specific acreage bases would be eliminated. Farmers would be free to plant crops based on market signals and what makes sense for their geographical and climatological area.

Specific Details

Eligibility: Farmers who participated in government supply management and farm programs in three of the last five years would be eligible.

The Contract: Farmers would sign a seven year contract with the United States Department of Agriculture to participate. The contract would require the continuance of conservation compliance plans already agreed to by the farmer.

The Payment: The current Congressional Budget Office baseline for Commodity Credit Corporation expenditure over the next 7 years (February, 1995 CBO) is approximately $?? billion. The Budget Resolution calls for a 7 year reduction in farm program spending of $13 billion. This will leave approximately $?? Billion to be spent by the CCC over the next 7 years.

156   APPENDIX 3

Seven Year Spending Path

1966 1997 1998 1999 2000 2001 2002

= $?? Billion

Numbers must be provided by CBO

The farmer's payment will be a prorated percentage of his historical payments, so that the program(s) will spend no more than the allotted amount in any one year. The payment to be made each October 15 will be according to a seven year contract with the farmer that spells out exactly the amount of his payment. The farmer will receive his prorated payment regardless of the level of commodity prices.

Advantages of This Type of Program

In general, this type of program would be a major departure from the past. Instead of the 1995 Farm bill offering the same old cobbled together 1930's era farm programs with bigger cuts to comply with budget mandates this would be a bold approach that tells the farmer he must transition to planting for world markets. Under the current programs, when prices improve either through market demand or bad crops farmers' deficiency payments are reduced. Under this program, the payment would be fixed, albeit a declining amount that the farmer could count on regardless of what happens to prices. It would remove farm programs and agricultural policy from the annual budget process because the farmer would have a contract with the Federal government for the next seven years.

This type of program would get the USDA out of the business of telling farmers how much to plant. This type of program would answer farm program critics' complaint that we should free up farmers to plant. It would eliminate supply management as the basis for farm programs and over time force production agriculture to allocate its resources into crops and areas where it makes the most sense economically to grow certain crops.

This type of program would be environmentally friendly. Farmers would not have to plant based upon their USDA acreage bases. This means that they could rotate crops more freely to control weed and other pest problems.

FLEXIBILITY:

++Gives maximum flexibility to farmers to plant for the market

++Eliminates ARPs and effectively establishes a Whole Farm Base

++Rewards farmers who plant in response to market signals

SIMPLICITY:

++Eliminates much of the paperwork at CFSA

++Reduces the farmers' need to certify acreage annually

++Should greatly reduce the farmers' trips to the CFSA office

++Eliminates the paperwork nightmare of acreage bases

++Eliminates the 5 month and 12 month deficiency payment calculation

CERTAINTY:

++Farmers and bankers know exactly payments to be received in the next seven years

++Answers the budget critics that CCC can be a "runaway" entitlement

++Eliminates the potential for payback of deficiency payments

++Eliminates the convoluted nature of the current program

Disadvantages:

++Farm programs would essentially be a "capped" entitlement

++Moves farm program payments into the category of "welfare Payments"

++Implies that the Federal involvement may be minimal or non-existent in the future

++In periods of low prices, the payment to farmers would remain the same

Appendix 4
Chronology for the 1996 Farm Act

May 1994, National Grain and Feed Association Foundation-supported study is released ("Large-Scale Land Idling Has Retarded Growth of U.S. Agriculture").

November 1994, Election of a Republican Congress, Congressman Pat Roberts and Sen. Richard G. Lugar become chairmen of the House and Senate Agriculture Committees.

December 1994, Senator Lugar circulates his questions on farm programs.

January 1995, Senator Lugar presents his plan to cut target prices 3 percent per year.

January 1995, The anonymous three-page paper and related one-page paper are circulated.

May 1995, The administration presents its plan in the "Blue Book."

May 1995, The House aggies boycott the leadership's budget conference, Roberts wins increases in the amount for commodity programs and a leadership promise to revisit budgets if agriculture falls on hard times.

July 1995, Roberts wins turf battle with Congressman Joe Skeen and the House Agriculture Appropriations Subcommittee.

July 1995, The anonymous three-page paper outlining the Freedom to Farm (FTF) proposal appears in the House.

July 1995, Sen. Patrick J. Leahy takes the position "if farm programs become the enemy of the hungry and the environment, I will not support them."

August 1995, Congressmen Roberts and Bill Barrett introduce an FTF bill in the House.

August 1995, Farm program cost estimates are reduced from February 1995 estimate of $14.4 billion to $6.9 billion for FY1997.

September 20, 1995, Roberts's FTF fails to pass in a late-night House Agriculture Committee session.

September 28, 1995, Senate Agriculture Committee passes a commodity title within the structure of current farm programs.

October 26, 1995, House budget reconciliation passes with FTF inserted by the House leadership.

November 15, 1995, FTF commodity title is reported out of House-Senate conference committee with major concessions to cotton and rice.

December 6, 1995, President vetoes budget reconciliation bill.

December 31, 1995, 1990 Farm Act expires.

January 8, 1996, Roberts introduces FTF in the House.

January 31–February 5, 1995, A Lugar/Leahy deal is in process.

February 8, 1995, The Senate passes a farm bill with FTF, conservation, nutrition, and the northeast dairy compact.

February 28, 1996, House passes an FTF commodity title.

March 21, 1996, Successful House-Senate conference completed on what becomes the 1996 Farm Bill.

April 4, 1996, President signs the Federal Agriculture Improvement and Reform Act of 1996—the 1996 Farm Act.

Appendix 5

CORN PRICES AND 1996 FARM ACT MILESTONES

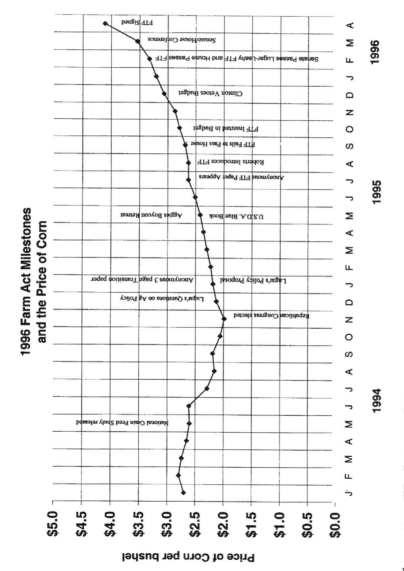

Notes

Chapter 1

1. Abel, Daft, & Earley, *Large-Scale Land Idling Has Retarded Growth of U.S. Agriculture*, prepared for the National Grain and Feed Foundation (Alexandria, Va.: Abel, Daft, & Earley, May 1994).

2. *What Happens to Farm Income* (Washington, D.C.: U.S. Agriculture 20/20, n.d.).

3. Sen. Richard G. Lugar, Washington, D.C., 22 December 1994, letter soliciting responses to questions regarding farm programs.

4. Sen. Richard G. Lugar, "Draft Questions for Comprehensive Senate Agriculture Committee Hearings on the 1995 Farm Bill," prepared by the Republican staff of the Senate Committee on Agriculture, Nutrition, and Forestry, Washington, D.C., 9 December 1994.

5. Sen. Richard G. Lugar to the Honorable Pete V. Domenici, 1 December 1994.

6. Sen. Richard G. Lugar, *Senator Richard Lugar Discusses Farm Program Cuts*, transcription of interview of Senator Lugar by Linda Wertheimer on National Public Radio (Journal Graphics Broadcast Database, Segment Number 08, Show Number 1726, 13 January 1995).

7. "Possible Approach to Major Withdrawal of Federal U.S. Government Farm and Farm Product Regulations and Programs Designed to Restrain Production, Restrict Market Supplies, and Increase Prices" [on or about January 1995].

8. "The Freedom to Farm Act of 1995: A Seven Year Contract with Production Agriculture," summer 1995.

9. Mike Espy and Alice M. Rivlin to the Honorable Patrick J. Leahy, Washington, D.C., 30 September 1994.

10. President to the Honorable Patrick J. Leahy, Washington, D.C., 30 September 1994.

11. Alice M. Rivlin, Big Choices: memorandum for handout and retrieval in meeting, Washington, D.C., 3 October 1994.

12. Sen. Richard G. Lugar, "Statement of U.S. Sen. Richard G. Lugar, Chairman Senate Committee on Agriculture, Nutrition, and Forestry," Washington, D.C., 16 February 1995.

13. Sen. Kent Conrad, "Statement of Senator Kent Conrad: Senate Committee on the Budget," Washington, D.C., 16 February 1995.

14. Milton C. Hallberg, Robert G. F. Spitze, and Daryll E. Ray, eds., *Food, Agriculture, and Rural Policy into the Twenty-First Century* (Boulder, Westview Press, 1994).

15. William Lin, Peter Riley, and Sam Evans, *Feed Grains: Background for 1995 Farm Legislation*, Agricultural Economics Report, no. 714. Other reports authored by Economic Research Service (ERS) economists include Agricultural Economics Report nos. 705–13 and 715–16. These reports focus on dairy, cotton, federal marketing orders and federal research and promotion programs, honey, tobacco, peanuts, sugar, wheat, rice, oilseeds, and agricultural export programs.

16. National Center for Food and Agricultural Policy, *Report of the Working Group on U.S. Farm Price and Income Stability*, NCFAP-95WG-02 (Washington, D.C.: National Center for Food and Agricultural Policy, March 1995).

17. U.S. General Accounting Office to the Honorable Richard G. Lugar and the Honorable Patrick J. Leahy, GAO/RCED-95-93R, Farm Bill Issues, Washington, D.C., 21 February 1995. Chuck Conner and Ed Barron to staff members of U.S. Senate Committee on Agriculture, Nutrition, and Forestry and legislative assistants to members of committee, Washington, D.C., 3 March 1995, with attachment *Abstracts of Responses to Senators Leahy and Lugar July 18, 1994 Letter Asking for Suggested Changes in Farm Legislation*, Washington, D.C., 1 March 1995.

CHAPTER 2

1. David Maraniss and Michael Weisskopf, "Aggies and the Road Gang Crunch the Numbers," *Washington Post*, 26 May 1995, provided by CompuServe Executive News Service.

2. *Congressional Record*, 104th Cong., 1st sess., 1995, 141, pt. 83:H5309.

3. Congressman Joe Skeen, chairman of the Agriculture Appropriations Subcommittee, in introducing the debate on H.R. 1976, Agriculture, Rural Development, Food and Drug Administration, and Related Agencies Appropriations Act, 1996, *Congressional Record*, 104th Cong., 1st sess., 1995, 141, pt. 117:H7240.

4. Recorded vote during debate on H.R. 1976, Agriculture, Rural Development, Food and Drug Administration, and Related Agencies Appropriations Act, 1996, *Congressional Record*, 104th Cong., 1st sess., 1995, 141, pt. 117:H7249.

5. Congressman Dan Miller, R-Fla., during debate on H.R. 1976, Agriculture, Rural Development, Food and Drug Administration, and Related Agencies Appropriations Act, 1996, *Congressional Record*, 104th Cong., 1st sess., 1995, 141, pt. 117:H7244.

6. Congresswoman Nita M. Lowey during debate on H.R. 1976, Agriculture, Rural Development, Food and Drug Administration, and Related Agencies Appropriations Act, 1996, *Congressional Record*, 104th Cong., 1st sess., 1995, 141, pt. 117:H7243.

7. "The Freedom to Farm Act of 1995: A Seven Year Contract with Production Agriculture," summer 1995.

8. Charles Abbot, "Cap on Spending Goes with Looser U.S. Farm Rules," Reuters Ltd., 24 July 1995, provided by CompuServe Executive News Service.

CHAPTER 3

1. Andy Morton, "The Making of the 1996 Farm Act as Seen from the Senate," in Otto C. Doering III and Lyle Schertz, eds., *1996 Farm Legislation: A Synopsis of a Pre-Conference at the 1996 Annual Meeting of the American Agricultural Economics Association* (Ames, Iowa: American Agricultural Economics Association, 1996).
2. "Family Farms Should Be Focus of Farm Bill—Dorgan," Reuters Ltd., 21 April 1995, provided by CompuServe Executive News Service.
3. Sen. Thomas A. Daschle, "Statement of Senate Democratic Leader Tom Daschle: Targeted Marketing Loan Proposal," U.S. Senate, Office of the Democratic Leader, Washington, D.C., 11 August 1995.
4. "U.S. Lawmaker—Dump Farm Program Before Food Aid," Reuters Ltd., 27 July 1995, provided by CompuServe Executive News Service.

CHAPTER 4

1. U.S. Department of Agriculture, *1995 Farm Bill: Guidance of the Administration* (Washington, D.C.: USDA, n.d.; 10 May 1995 on related press release).
2. "U.S. Senate Ag Chair Lugar Calls Farm Plan Irrelevant," Reuters Ltd., 10 May 1995, provided by CompuServe Executive News Service.
3. Guy Gugliotta, "Administration Issues Farm Plan; Agriculture Secretary Would Preserve Nutrition, Subsidy Programs," *Washington Post*, 11 May 1995, provided by CompuServe Executive News Service.
4. "Clinton Administration Unveils Farm Bill Proposals," Reuters Ltd., 10 May 1995, provided by CompuServe Executive News Service.
5. "Glickman Hints at More Unpaid Acres," *SCI Policy Report*, 4 May 1995.
6. "USDA to Rely on Farmers' Honor in Subsidy Cut-off," Reuters Ltd., 11 May 1995, provided by CompuServe Executive News Service.
7. "Clinton Calls U.S. Conservation Reserve a Success," Reuters Ltd., 1 June 1995, provided by CompuServe Executive News Service.
8. Robert Greene, "Glickman—Farm Budget," Associated Press, 14 June 1995, provided by CompuServe Executive News Service.
9. "Clinton Eyes $4 Billion in Farm Cuts—U.S. Senators," Reuters Ltd., 14 June 1995, provided by CompuServe Executive News Service.
10. Charles Abbott, "U.S. Groups Split on Crop Flexibility in Farm Bill," Reuters Ltd., 19 June 1995, provided by CompuServe Executive News Service.
11. "Clinton Shifts Position on Food Program Cuts," Reuters Ltd., 14 June 1995, provided by CompuServe Executive News Service.
12. "USDA's Glickman Backs Farm-Welfare Coalition," Reuters Ltd., 19 June 1995, provided by CompuServe Executive News Service.
13. Judith Havemann, "Packwood Backs State Food Stamp Plan; Senate

Chairman Favors Welfare Reform Separate from Budget Bill," *Washington Post*, 27 April 1995, provided by CompuServe Executive News Service.

14. Charles Abbot, "Glickman Urges Clinton Veto of Food Stamp Bill," Reuters Ltd., 23 May 1995, provided by CompuServe Executive News Service.

15. Charles Abbott, "USDA Urges Veto of Shift in Food Aid to States," Reuters Ltd., 23 May 1995, provided by CompuServe Executive News Service.

16. Judith Havemann, "Lugar Switches on School Lunches, Food Stamps; Senate Agriculture Chairman No Longer Favors Turning the Federal Programs Over to States," *Washington Post*, 10 June 1995, provided by CompuServe Executive News Service.

17. "U.S. Senate Leans Against Food Stamp Block Grants," Reuters Ltd., 28 July 1995, provided by CompuServe Executive News Service.

CHAPTER 5

1. United States Department of Agriculture, *Impact of the Republican Budget Cuts on Rural America: A State-by-State Analysis* (Washington, D.C.: USDA, 11 October 1995).

2. "Zimmer Intrigued By 'Freedom to Farm' Proposal," *SCI Policy Report*, 2 August 1995.

3. "Details of Gunderson's Dairy Reform Plan to Be Released Today," *SCI Morning Comments*, 3 August 1995.

4. "Administration Forecasts Lower Farm Program Costs," *SCI Policy Report*, 3 August 1995.

5. "Many Questions Remain on 'Freedom to Farm Act,'" *SCI Policy Report*, 7 August 1995.

6. "Rep. Roberts', Others' Comments on Farm Act (FFA)," *SCI Morning Comments*, 7 August 1995.

7. "Crop Insurance Fight Brewing Over Roberts' Bill," *The Food & Fiber Letter*, 7 August 1995.

8. "House Ag Cmte. Document Outlines Farm Bill Timeline," *SCI Morning Comments*, 8 August 1995.

9. For a brief comparison of the three most prominent proposals in late summer 1995 see "Washington's 'Dullest Month' Yields Feast of Competing Farm Bill Plans," *Webster Agricultural Letter* 2, no.16 (15 August 1995).

10. "Another Farm Bill Alternative Emerges," *SCI Policy Report*, 11 August 1995.

11. Ibid.

12. "Sen. Dems Offer Two-Tiered Loan Farm Bill Proposal," *SCI Morning Comments*, 14 August 1995.

13. Sen. Thomas A. Daschle, "Statement of Senate Democratic Leader Tom Daschle: Targeted Marketing Loan Proposal," U.S. Senate, Office of the Democratic Leader, Washington, D.C., 11 August 1995.

14. Andy Morton, "The Making of the 1996 Farm Act as Seen from the Senate," in Otto C. Doering III and Lyle Schertz, eds., *1996 Farm Legislation: A Synopsis of a Pre- Conference at the 1996 Annual Meeting of the American Agricultural Economics Association* (Ames, Iowa: American Agricultural Economics Association, 1996).

Chapter 6

1. Technically, the committees were required only to achieve outlay savings for all programs under their jurisdiction, and they were free to assign the savings among the programs as they thought was appropriate. Nonetheless, the deliberations of the Budget Committees focused on specific numbers for Budget Function Area 350, which is the budget area that includes the commodity programs as well as other USDA program areas like research. Consequently, it would have been difficult for the committees to ignore the particular savings that the Budget Committees had identified for the 350 function by, say, proposing even deeper cuts in the food programs than were identified in the Budget Committee discussions.

2. Sara Hebel, "Report on Markup of H.R. 2491: Seven-Year Balanced Budget Reconciliation Act of 1995 in the House Committee on Agriculture," Legi-Slate News Service, 20 September 1995.

3. Technically, the entitlement was associated with land, that is, if the owners and operators of land parcels had received farm commodity payments in past years, the current owners and operators were eligible to receive the payments.

4. Computation of a five-year olympic average in this case involves discarding the highest and the lowest prices among the five annual prices and then calculating the mean of the other three prices.

5. Sara Hebel, "House Agriculture Revolt Puts GOP Reform Out to Pasture," Legi-Slate News Service, 21 September 1995.

6. Sara Hebel, "Report on Markup of H.R. 2491: Seven-Year Balanced Budget Reconciliation Act of 1995 in the House Committee on Agriculture," Legi-Slate News Service, 20 September 1995.

7. Sara Hebel, "Report on Markup of Draft (unnumbered). House Agriculture Panel Deadlocks on Farm Reforms in the House Committee on Agriculture," Legi-Slate News Service, 20 September 1995.

8. Sara Hebel, "Report on Markup of Draft (unnumbered). House Ag Committee Gives Up on Reconciliation Plan in the House Committee on Agriculture," Legi-Slate News Service, 27 September 1995.

9. Congressman John R. Kasich, Chairman, House Committee on the Budget to the Honorable Pat Roberts, Washington, D.C., 14 September 1995.

10. Congressmen Richard K. Armey, Newt Gingrich, and Tom DeLay to the Honorable Pat Roberts, Chairman, House Committee on Agriculture, 14 September 1995.

11. "The Farm Bill Debate Is in a State of Flux," *SCI Policy Report*, 3 October 1995.

12. Ibid.

13. "'Gang of Four' Threatens to Vote Against Reconciliation," *SCI Policy Report*, 17 October 1995.

14. "Farm Language of House Reconciliation Still Under Fire," *SCI Policy Report*, 23 October 1995.

15. "House Members Write Gingrich, Opposing Freedom to Farm," *SCI Morning Comments*, 25 October 1995.

16. "GOP House Leadership Prevails on Farm Issues," *SCI Policy Report*, 26 October 1995.

168   NOTES

17. "Report on Markup of Draft (unnumbered): Senate Agriculture Panel Breaks Deadlock, Passes Farm Reforms in the Senate Committee on Agriculture, Nutrition, and Forestry," Legi-Slate News Service, 28 September 1995. "Senate Panel Agrees on Farm Cuts," United Press International, 28 September 1995, provided by CompuServe Executive News Service. Robert Greene, "Farm Bill," Associated Press, 28 September 1995, provided by CompuServe Executive News Service.

18. "Senate Passes Reconciliation; Contentious Conference Ahead," *SCI Policy Report*, 30 October 1995.

19. Bruce Gardner, "Farm Revenues and Net Income Under House and Senate Farm Legislation in the 1995 Budget Reconciliation Act" (paper presented at USDA Farm Bill Forum, 31 October 1995).

20. "Staff Level Talks Continue on Farm Reconciliation," *SCI Policy Report*, 7 November 1995.

21. "Roberts Says Farm Reconciliation Has Been Reached," *SCI Policy Report*, 10 November 1995.

22. Senate Agriculture Committee and House Agriculture Committee news release, "Conferees Reach Agreement on Agriculture Section of Balanced Budget Bill; Compromise Meets Mandate for Reform While Preserving Federal Safety Net for Farmers," 15 November 1995.

23. "Highlights of Changes to Ag Reconciliation Package," *SCI Morning Comments*, 16 November 1995.

24. House Agriculture Committee news release, "Roberts Announces Estimated Market Transition Payments; Farmers Won't Have to Pay Checks Back to the Government for 1995 Advance Payments," 21 November 1995.

25. President to the House of Representatives, the White House, Office of the Press Secretary, 6 December 1995, veto of H.R. 2491 message and attachment: "H.R. 2491: Objectionable Provisions."

CHAPTER 7

1. "U.S. Farm Bill Deadline Looms, Extension Possibility," Reuters Ltd., 30 November 1995, provided by CompuServe Executive News Service.

CHAPTER 8

1. "Glickman Opposed to 60-Day Early-Out for CRP," *SCI Morning Comments*, 1 December 1995.

2. "Budget Update," *SCI Morning Comments*, 8 December 1995.

3. "Democrats Still Pushing for Higher Loan Rates," Knight Ridder, 15 December 1995, provided by CompuServe Executive News Service.

4. "Administration Could Tip the Scales, Extend Farm Programs," *Webster Agricultural Letter* 3, no. 2 (15 January 1996).

5. "Lugar Urges Guaranteed Payment Concept Be Maintained as Ag Budget Component," *SCI Morning Comments*, 15 December 1995.

6. "Glickman Cuts Trip Short as Negotiations Intensify," *SCI Policy Report*, 30 January 1996.

7. "Summit Called to Break Farm Policy Logjam," *SCI Policy Report,* 23 January 1996.

8. Ibid.

9. "Summit Offers No Solutions but Work to Continue," *SCI Policy Report,* 24 January 1996.

10. "Summit Called to Break Farm Policy Logjam," *SCI Policy Report,* 23 January 1996.

11. "Roberts Foregoes CR, Seeks a Farm Deal with Democrats," *SCI Policy Report,* 26 January 1996.

12. Congressional Budget Office, CBO memorandum: The Economic and Budget Outlook: December 1995 Update, Washington, D.C., n.d. Congressional Budget Office, *Commodity Credit Corporation: November 1995 CBO Baseline* (Washington, D.C.: CBO, 5 December 1995).

13. "National Corn Growers Assn. Pushes for Action on Budget," *SCI Morning Comments,* 8 December 1995.

14. "House Committee Releases List of Groups Backing Freedom to Farm," *SCI Morning Comments,* 19 December 1995.

15. "American Soybean Association Backs Daschle Farm Bill Plan," *SCI Morning Comments,* 18 December 1995.

16. "Economists Pressure White House to Accept Freedom to Farm," *SCI Policy Report,* 18 December 1995.

17. "AFBF Urges Congress to Resolve Budget Stalemate," *SCI Policy Report,* 16 January 1996.

CHAPTER 9

1. Sara Hebel, "Agriculture Meeting Yields No Agreement on Farm Policy," Legi-Slate News Service, 23 January 1996.

2. "Dole Lauds U.S. Ag Reforms, Daschle Sees Filibuster," Reuters Ltd., 23 January 1996, provided by CompuServe Executive News Service.

3. Sara Hebel, "Agriculture Meeting Yields No Agreement on Farm Policy," Legi-Slate News Service, 23 January 1996.

4. "House, Senate May Differ on U.S. Farm Bill Contents," Reuters Ltd., 29 January 1996, provided by CompuServe Executive News Service.

5. "Chance to Pass U.S. Farm Bill This Week—Sen. Daschle," Reuters Ltd., 29 January 1996, provided by CompuServe Executive News Service.

6. "Senate Likely to See Three U.S. Farm Bill Options," Reuters Ltd., 30 January 1996, provided by CompuServe Executive News Service.

7. Robert Greene, "Farm Bill," Associated Press, 30 January 1996, provided by CompuServe Executive News Service.

8. "Daschle Sees Hope for Compromise Farm Bill," Reuters Ltd., 30 January 1996, provided by CompuServe Executive News Service.

9. "Farm Bill Delayed: Democratic Opposition to Bill," Dow Jones News, 2 February 1996, provided by CompuServe Executive News Service.

10. "Senate to Begin Voting on Farm Bill," Dow Jones News, 1 February 1996, provided by CompuServe Executive News Service.

11. "Dole to Decide by 6 P.M. If Senate Will Vote on Farm Bill," Dow Jones

News, 1 February 1996, provided by CompuServe Executive News Service.

12. *Congressional Record*, 104th Cong., 2nd. sess, 1996, 142, pt. 14:S683.

13. "U.S. Senate Leaders Seek Farm-Subsidy Compromise," Reuters Ltd., 1 February 1996, provided by CompuServe Executive News Service.

14. "New 3-year U.S. Farm Bill Being Written in Senate," Reuters Ltd., 1 February 1996, provided by CompuServe Executive News Service.

15. "Dole to Decide by 6 P.M. If Senate Will Vote on Farm Bill," Dow Jones News, 1 February 1996, provided by CompuServe Executive News Service.

16. "Dole Delays Senate Cloture Vote on Compromise Farm Bill," Dow Jones News, 1 February 1996, provided by CompuServe Executive News Service.

17. Ibid.

18. "U.S. Senate Puts Off Farm Bill Votes to Tuesday," Reuters Ltd., 1 February 1996, provided by CompuServe Executive News Service.

19. "U.S. Farm Bill 3, 4, or 5 Years, Split Payment—Lugar," Reuters Ltd., 1 February 1996, provided by CompuServe Executive News Service.

20. "U.S. Farm Talks Moving 'in Right Direction'—Glickman," Reuters Ltd., 2 February 1996, provided by CompuServe Executive News Service.

21. "Legislation Would Raise Price of Milk—CBS," Reuters Ltd., 2 February 1996, provided by CompuServe Executive News Service.

22. *Congressional Record*, 104th Cong., 2d. sess., 1996, 142, pt. 16:S882.

23. "Lugar Sees Close Senate Vote on Allowing Debate," Knight Ridder, 6 February 1996, provided by CompuServe Executive News Service.

24. *Congressional Record*, 104th Cong., 2d. sess., 1996, 142, pt. 16:S907.

25. Sara Hebel, "One Vote Delays Senate Farm Reforms; Dole Seeks Wednesday Vote," Legi-Slate News Service, 23 January 1996.

26. "Senate Returns for Still Another Try at Farm Bill," *SCI Policy* Report, 7 February 1996.

27. *Congressional Record*, 104th Cong., 2d. sess., 1996, 142, pt. 17:S1043-1049 and 1141.

28. *Congressional Record*, 104th Cong., 2d. sess., 1996, 142, pt. 17:S1049 and 1052.

29. *Congressional Record*, 104th Cong., 2d. sess., 1996, 142, pt. 17:S1040.

30. *Congressional Record*, 104th Cong., 2d. sess., 1996, 142, pt. 17:S1052.

31. *Congressional Record*, 104th Cong., 2d. sess., 1996, 142, pt. 17:S1054.

32. *Congressional Record*, 104th Cong., 2d. sess., 1996, 142, pt. 17:S1059.

33. "Senate Approves Farm Bill," United Press International, 7 February 1996, provided by CompuServe Executive News Service.

34. "Congress This Week," SCI Morning Comments, 26 February 1996.

35. Robert Greene, "Farm Bill," Associated Press, 8 February 1996, provided by CompuServe Executive News Service.

36. Mike Glover, "Farm Bill," Associated Press, 9 February 1996, provided by CompuServe Executive News Service.

37. Philip Brasher, "Daschle's Defeat," Associated Press, 10 February 1996, provided by CompuServe Executive News Service.

38. "Clinton Outlines Concerns On Senate-Passed Farm Bill," *SCI Policy Report*, 13 February 1996.

CHAPTER 10

1. "U.S. House Ag Chair Will Eye Farm Reform Changes," Reuters Ltd., 25 January 1996, provided by CompuServe Executive News Service.
2. "Market Transition Act Okayed on Party Lines, Faces Veto Threat," *SCI Policy Report*, 31 January 1996.
3. "Two-year U.S. Farm Bill Suggested as Alternative," Reuters Ltd., 24 January 1996, provided by CompuServe Executive News Service.
4. "No Agreement on New Farm Bill at Summit Meeting," Reuters Ltd., 23 January 1996, provided by CompuServe Executive News Service.
5. Robert Greene, "Dairy Overhaul," Associated Press, 25 January 1996, provided by CompuServe Executive News Service.
6. "USDA Analysis: House Dairy Plan Boosts Nutrition Program Costs," *SCI Morning Comments*, 16 February 1996.
7. Sara Hebel, "Report on Markup of H.R. 2854, Federal Agriculture Improvement and Reform Act of 1996 (Farm Bill) in the House Committee on Agriculture," Legi-Slate News Service, 30 January 1996.
8. Robert Greene, "Farm Bill," Associated Press, 30 January 1996, provided by CompuServe Executive News Service.
9. Sara Hebel, "Report on Markup of H.R. 2854, Federal Agriculture Improvement and Reform Act of 1996 (Farm Bill) in the House Committee on Agriculture," Legi-Slate News Service, 30 January 1996.
10. Ibid.
11. "Farm Compromise Efforts Falter on Both Sides of the Hill," *SCI Policy Report*, 2 February 1996.
12. "House Appropriation Leaders Warn on Farm Bill Provisions," *SCI Morning Comments*, 21 February 1996.
13. "House Ag Committee Chairman Introduces Farm Bill II," *SCI Policy Report*, 28 February 1996.
14. "House Ag Committee Chairman Roberts Unveils Farm Bill II," *SCI Morning Comments*, 28 February 1996.
15. Sara Hebel, "Rep. Roberts Offers Farm Bill II: Food Fight on House Floor Likely," Legi-Slate News Service, 27 February 1996.
16. See *Congressional Record,* 104th Cong., 2d. sess., 1996, 142, pt 25:H1474 and H1478 for how Congressman Charlie Rose, D-N.C., used the Cuban downing of the planes in arguing against the Miller-Schumer amendment, which would have changed the sugar program greatly by lowering the loan rates, and how Congressman Charles E. Schumer, D-N.Y., rebutted Rose's claims.
17. Sara Hebel, "House Takes Up Farm Bill, Expected to Add Conservation Compromise," Legi-Slate News Service, 28 February 1996.
18. *Congressional Record*, 104th Cong., 2d. sess., 1996, 142, pt. 26: H1509.

172   NOTES

CHAPTER 11

1. Sen. Patrick J. Leahy, "Statement of Senator Patrick J. Leahy," Washington, D.C., 20 March 1996.
2. Bryan Just and Linwood Hoffman, "Permanent Price Support Authority," in Frederick J. Nelson and Lyle P. Schertz, eds., *Provisions of the Federal Agriculture Improvement and Reform Act of 1996* (Washington, D.C.: USDA, AIB No. 729, September 1996), pp. 24–26.
3. Public Law 104–127, sec. 171(e).
4. Richard Stillman, "Dairy," and Bryan Just and Linwood Hofman, "Permanent Price Support Authority," in Frederick J. Nelson and Lyle P. Schertz, eds., *Provisions of the Federal Agriculture Improvement and Reform Act of 1996* (Washington, D.C.: USDA, AIB No. 729, September 1996), p. 15.
5. Scott Sanford, "Peanuts," and Bryan Just and Linwood Hofman, "Permanent Price Support Authority," in Frederick J. Nelson and Lyle P. Schertz, eds., *Provisions of the Federal Agriculture Improvement and Reform Act of 1996* (Washington, D.C.: USDA, AIB No. 729, September 1996), pp. 17–19.
6. President, "Statement by the President on the Farm Bill Signing," The White House, Office of the Press Secretary, 4 April 1996.
7. Dan Glickman, "Glickman Pledges Swift Implementation of New Farm Law," release no. 0173.96, Washington, D.C.: USDA, 4 April 1996.
8. "Glickman Sees Possibility of Farm Bill Changes," Reuters Ltd., 4 April 1996.

CHAPTER 13

1. Chester C. Davis, "The Development of Agricultural Policy Since the End of the World War," in *Farmers in a Changing World* (Washington, D.C.: GPO, 1940), p. 325.
2. Howard R. Tolley, "Some Essentials of a Good Agricultural Policy," in *Farmers in a Changing World* (Washington, D.C.: GPO, 1940), p. 1169.
3. Some sources that provide a more complete picture of traditional American farm programs include: Willard W. Cochrane, *The Development of American Agriculture*, 2d ed. (Minneapolis: University of Minnesota Press, 1993); Congressional Research Service, *Farm Bill Issues Overview* (Washington, D.C.: Library of Congress, 25 March 1996); U.S. Department of Agriculture, Economic Research Service, *History of Agricultural Price-Support and Adjustment Programs, 1933–1984*, Agriculture Information Bulletin No. 485; I. Roberts, Neil Andrews, Suzanne Doyle, Rick Cannan, Peter Connell, and Ahmed Hafi, *U.S. Farm Bill 1995; U.S. Agricultural Policies on the Eve of the 1995 Farm Bill* (Canberra: Australian Bureau of Agricultural and Resource Economics, 1995); U.S. Department of Agriculture, *Farmers in a Changing World* (Washington, D.C.: GPO, 1940); and U.S. Department of Agriculture, *A Time to Choose: Summary Report on the Structure of Agriculture* (Washington, D.C.: USDA, 1981).

BIBLIOGRAPHY

Abbott, Charles. "Glickman Urges Clinton Veto of Food Stamp Bill." Reuters Ltd., 23 May 1995. Provided by CompuServe Executive News Service.

Abbott, Charles. "USDA Urges Veto of Shift in Food Aid to States." Reuters Ltd., 23 May 1995. Provided by CompuServe Executive News Service.

Abbott, Charles. "U.S. Groups Split on Crop Flexibility in Farm Bill." Reuters Ltd., 19 June 1995. Provided by CompuServe Executive News Service.

Abbot, Charles. "Cap on Spending Goes with Looser U.S. Farm Rules." Reuters Ltd., 24 July 1995. Provided by CompuServe Executive News Service.

Abel, Daft, & Earley. *Large-Scale Land Idling Has Retarded Growth of U.S. Agriculture*. Prepared for the National Grain and Feed Foundation. Alexandria, Va.: Abel, Daft & Earley, May 1994.

Armey, Rep. Richard K., Rep. Newt Gingrich, and Rep. Tom DeLay. Letter to the Honorable Pat Roberts, Chairman, House Committee on Agriculture. 14 September 1995.

Brasher, Philip. "Daschle's Defeat." Associated Press, 10 February 1996. Provided by CompuServe Executive News Service.

Cochrane, Willard W. *The Development of American Agriculture,* 2d ed. Minneapolis: University of Minnesota Press, 1993.

CompuServe Executive News Service. 1995–96.

Congressional Budget Office. CBO memorandum: The Economic and Budget Outlook: December 1995 Update. Washington, D.C., n.d.

Congressional Budget Office. *Commodity Credit Corporation: November 1995 CBO Baseline.* Washington, D.C.: CBO, 5 December 1995.

Congressional Record. 1994–96. Washington, D.C.

Congressional Research Service. *Farm Bill Issues Overview.* Washington, D.C.: Library of Congress, 25 March 1996.

Conner, Chuck, and Ed Barron. Memorandum to staff members of U.S. Senate Committee on Agriculture, Nutrition, and Forestry and legislative assistants to members of committee. Washington, D.C., 3 March 1995. With attachment *Abstracts of Responses to Senators Leahy and Lugar July 18, 1994 Letter Asking for Suggested Changes in Farm Legislation.* Washington, D.C., March 1, 1995.

Conrad, Sen. Kent. "Statement of Senator Kent Conrad: Senate Committee on the Budget." Washington, D.C., 16 February 1995.

Daschle, Sen. Thomas A. "Statement of Senate Democratic Leader Tom Daschle: Targeted Marketing Loan Proposal." U.S. Senate, Office of the Democratic Leader, Washington, D.C., 11 August 1995.

Davis, Chester C. "The Development of Agricultural Policy Since the End of the World War." In *Farmers in a Changing World*. Washington, D.C.: GPO, 1940.

Doering, Otto C. III, and Lyle Schertz, eds. *1996 Farm Legislation: A Synopsis of a Pre-Conference at the 1996 Annual Meeting of the American Agricultural Economics Association*. Ames, Iowa: American Agricultural Economics Association, 1996.

Espy, Mike, and Alice M. Rivlin. Letter to the Honorable Patrick J. Leahy. Washington, D.C., 30 September 1994.

Federal Agriculture Improvement And Reform Act of 1996. Public Law 104–127, 4 April 1996.

Food & Fiber Letter. April 1994–April 1996.

"The Freedom to Farm Act of 1995: A Seven Year Contract with Production Agriculture." Summer 1995.

Gardner, Bruce. "Farm Revenues and Net Income Under House and Senate Farm Legislation in the 1995 Budget Reconciliation Act." Paper presented at USDA Farm Bill Forum, 31 October 1995

Glickman, Dan. "Glickman Pledges Swift Implementation of New Farm Law." Release No. 0173.96. U.S. Department of Agriculture, Washington, D.C., 4 April 1996.

Greene, Robert. "Glickman—Farm Budget." Associated Press, 14 June 1995. Provided by CompuServe Executive News Service.

Greene, Robert. "Dairy Overhaul." Associated Press, 25 January 1996. Provided by CompuServe Executive News Service.

Greene, Robert. "Farm Bill." Associated Press, 30 January 1996. Provided by CompuServe Executive News Service.

Gugliotta, Guy. "Administration Issues Farm Plan; Agriculture Secretary Would Preserve Nutrition, Subsidy Programs." *Washington Post*, 11 May 1995. Provided by CompuServe Executive News Service.

Hallberg, Milton C., Robert G. F. Spitze, and Daryll E. Ray, eds. *Food, Agriculture, and Rural Policy into the Twenty-First Century*. Boulder: Westview Press, 1994.

Havemann, Judith. "Packwood Backs State Food Stamp Plan; Senate Chairman Favors Welfare Reform Separate from Budget Bill." *Washington Post*, 27 April 1995. Provided by CompuServe Executive News Service.

Havemann, Judith. "Lugar Switches on School Lunches, Food Stamps; Senate Agriculture Chairman No Longer Favors Turning the Federal Programs Over to States." *Washington Post*, 10 June 1995. Provided by CompuServe Executive News Service.

Hebel, Sara. "Report on Markup of Draft (unnumbered). House Agriculture Panel Deadlocks on Farm Reforms in the House Committee on Agriculture." Legi-Slate News Service, 20 September 1995.

Hebel, Sara. "Report on Markup of H.R. 2491: Seven-Year Balanced Budget Reconciliation Act of 1995 in the House Committee on Agriculture." Legi-Slate News Service, 20 September 1995.
Hebel, Sara. "House Agriculture Revolt Puts GOP Reform Out to Pasture." Legi-Slate News Service, 21 September 1995.
Hebel, Sara. "Report on Markup of Draft (unnumbered). House Ag Committee Gives Up on Reconciliation Plan in the House Committee on Agriculture." Legi-Slate News Service, 27 September 1995.
Hebel, Sara. "Report on Markup of H.R. 2854, Federal Agriculture Improvement and Reform Act of 1996 (Farm Bill) in the House Committee on Agriculture." Legi-Slate News Service, 30 January 1996.
Hebel, Sara. "Rep. Roberts Offers Farm Bill II: Food Fight on House Floor Likely." Legi-Slate News Service, 27 February 1996.
Hebel, Sara. "House Takes Up Farm Bill, Expected to Add Conservation Compromise." Legi-Slate News Service, 28 February 1996.
Kasich, Congressman John R. Letter to the Honorable Pat Roberts. Washington, D.C., 14 September 1995.
Leahy, Sen. Patrick J. "Statement of Senator Patrick J. Leahy." Washington, D.C., 20 March 1996.
Legi-Slate News Service. 1995–96. Washington, D.C.
Lugar, Sen. Richard G. Letter to Honorable Pete V. Domenici, 1 December 1994.
Lugar, Sen. Richard G. "Draft Questions for Comprehensive Senate Agriculture Committee Hearings on the 1995 Farm Bill." Prepared by the Republican staff of the Senate Committee on Agriculture, Nutrition, and Forestry. Washington, D.C., 9 December 1994.
Lugar, Sen. Richard G. Letter to farm groups soliciting responses to questions regarding farm programs. Washington, D.C., 22 December 1994.
Lugar, Sen. Richard G. *Senator Richard Lugar Discusses Farm Program Cuts*. Transcription of interview of Senator Lugar by Linda Wertheimer on National Public Radio. Journal Graphics Broadcast Database, Segment Number 08, Show Number 1726, 13 January 1995.
Lugar, Sen. Richard G. "Statement of U.S. Sen. Richard G. Lugar, Chairman Senate Committee on Agriculture, Nutrition, and Forestry." Washington, D.C., 16 February 1995.
Maraniss, David, and Michael Weisskopf. "Aggies and the Road Gang Crunch the Numbers." *Washington Post*, 26 May 1995. Provided by CompuServe Executive News Service.
National Center for Food and Agricultural Policy. *Report of the Working Group on U.S. Farm Price and Income Stability*. NCFAP-95WG-02. Washington, D.C.: National Center for Food and Agricultural Policy., March 1995
Nelson, Frederick J., and Lyle P. Schertz, eds. *Provisions of the Federal Agriculture Improvement and Reform Act of 1996*. Washington, D.C.: USDA, AIB No. 729, September 1996.
"Possible Approach to Major Withdrawal of Federal U.S. Government Farm and Farm Product Regulations and Programs Designed to Restrain Produc-

tion, Restrict Market Supplies, and Increase Prices." [On or about January 1995.]

President. Letter to the Honorable Patrick J. Leahy. Washington, D.C., 30 September 1994.

President. Letter to the House of Representatives. The White House, Office of the Press Secretary, 6 December 1995. Veto of H.R. 2491 message and attachment: "H.R. 2491: Objectionable Provisions."

President. "Statement by the President on the Farm Bill Signing." The White House, Office of the Press Secretary, 4 April 1996.

Rivlin, Alice M. Big Choices: Memorandum for handout and retrieval in meeting. Washington, D.C., 3 October 1994.

Roberts, I., Neil Andrews, Suzanne Doyle, Rick Cannan, Peter Connell, and Ahmed Hafi. *U.S. Farm Bill 1995; U.S. Agricultural Policies on the Eve of the 1995 Farm Bill.* Canberra: Australian Bureau of Agricultural and Resource Economics, 1995.

SCI Morning Comments. April 1994–April 1996. McLean, Va.

SCI Policy Report. April 1994–April 1996. McLean, Va.

Tolley, Howard R. "Some Essentials of a Good Agricultural Policy." In *Farmers in a Changing World.* Washington, D.C.: GPO, 1940.

U.S. Department of Agriculture. *Farmers in a Changing World.* Washington, D.C.: GPO, 1940.

U.S. Department of Agriculture. *A Time to Choose: Summary Report on the Structure of Agriculture.* Washington, D.C.: USDA, 1981.

U.S. Department of Agriculture. *1995 Farm Bill: Guidance of the Administration.* Washington, D.C., n.d.; 10 May 1995 on related press release.

U.S. Department of Agriculture. *Impact of the Republican Budget Cuts on Rural America: A State-by-State Analysis.* Washington, D.C.: USDA, 11 October 1995.

U.S. Department of Agriculture. Economic Research Service. *History of Agricultural Price-Support and Adjustment Programs, 1933–1984.* Agriculture Information Bulletin No. 485.

U.S. Department of Agriculture. Economic Research Service. *Background for 1995 Farm Legislation.* Agricultural Economic Reports AER 705–16.

U.S. General Accounting Office. Communication to the Honorable Richard G. Lugar and the Honorable Patrick J. Leahy. GAO/RCED-95-93R, Farm Bill Issues. Washington, D.C., 21 February 1995.

Washington Post. April 1994–April 1996.

Webster Agricultural Letter. April 1994–April 1996. Washington, D.C.

INDEX

Abel, Daft, & Earley, 4
Acreage Reduction Program (ARP), 155
  according to Blue Book, 46
  Agricultural Competitiveness Act (S. 1155) and, 58
  anonymous early 1995 farm program withdrawal and, 151
  Coalition for a Competitive Food and Agriculture System, 93–94
  Freedom to Farm Act of 1995 and, 64–65
  historical program aspects and, 135
  House-Senate conference agreement and, 74, 77
  Large-Scale Land Idling Has Retarded Growth of U.S. Agriculture and, 4–6
  Price and Income Stability Working Group and, 18–19
  Richard Lugar's questions for Senate hearings regarding, 142
  seven year program and, 157
  Thad Cochran's adjustments to, 43
Activities in Support of Farming Under Agricultural Act of 1949, 56
Administration. *See also* Clinton, Bill, Pres.
  $5 billion cut in farm program by, 87
  budget summit, 86–88
  food program blocking and, 49–50, 53
  marketing loans and, 88
Agricultural Acts
  of 1933, 139
  of 1938, 117–118, 134
  of 1949, 117–118
Agricultural Appropriations Subcommittees
  FY1996 balanced budget and, 28–31
  rival with Agriculture Committee, 27–31, 110
Agricultural colleges, 17–19
Agricultural Competitiveness Act (S. 1155)
  description, 57–58
  flexibility for operators in, 58–60
Agricultural economics. *See also* Commodity prices; Prices
  and future farm legislation, 127–129
  of WWII, 135–136
Agricultural Marketing Service, 141
Agricultural Research Service, 24–25
Agricultural Stabilization and Conservation Service (ASCS)
  anonymous early 1995 farm program withdrawal and, 151
  Richard Lugar's questions for Senate hearings regarding, 141
Agriculture Committees
  FY1996 balanced budget and, 28–31
  House negotiations and, 107
  record of farm legislative attitude, 122
  Republican control of, 3–4
  Senate approval of budget reconciliation, 71
Aid to Families with Dependent Children (AFDC), 51. *See also* Food assistance programs
Akaka, Daniel K. (Sen., D-Hawaii), 42
Alternative crops, 63
Amendments
  capping/termination of CRP contract, 108–109
  House Rules Committee decisions regarding, 110–112
  No. 1384, 101
  No. 1541, 102
  No. 3184, 102
  No. 3451, 102

177

Amendments (*Continued*)
  No. 3452, 101
  No. 3456, 102
  rejected during H.R. 2854 debate, 112–113
American Farm Bureau Federation (AFBF)
  endorsement of Freedom to Farm proposal, 92–93
  and Sen. Lugar's early agenda, 9–10
American Soybean Association, 93
Animal and Plant Inspection Service, 141
Appropriations
  1995 Republican delay of, 80
  budget resolutions and, 23–26
  FY1994/95 honey loans, 28
  FY1996 and, 25–26, 31–33
Armey, Richard K. (Rep., R-Tex.)
  1996 budgeting and, 29–30
  Freedom to Farm proposal and, 67–68
  as House majority leader, 7–8

Baize, John, 14
Baker, Richard H. (Rep., R-La.), 66
Balanced budget
  1995 administration's approach to, 45
  Agriculture Appropriations Subcommittee/Agriculture Committee rivalry and, 28–31
  Blue Book and, 47–49
  budget reconciliation bill and, 61–62
  entitlement programs and, 21–26
  food assistance programs and, 49–50
  FY1996 appropriation and, 25–26
  Republican-initiated budget cuts and, 22–23
  through $5 billion cut in farm program, 87
Barrett, Bill (Rep., R-Nebr.)
  Freedom to Farm Act of 1995 and, 36
  Freedom to Farm proposal and, 56–57
Barron, Ed, 103
Base acreage
  according to Emerson-Combest proposal, H.R. 2330, 63
  Agricultural Competitiveness Act (S. 1155) and, 59
  Freedom to Farm Act and, 137, 153
  two-tier marketing loans and, 58
Baucus, Max (Sen., D-Mont.), 41
Beltsville Agricultural Research Center, 32–33
Benson, Ezra Taft, 137
Big Choices memorandum, 15–16
Bishop, Sanford D., Jr. (Rep., D-Ga.), 108
Blue Book (USDA), 45–47, 159
Boehlert, Sherwood L. (Rep., R-N.Y.), 113
Boschwitz, Rudy (Sen., R-Minn.), 59
Brannan plan of 1948, 137
Breaux, John B. (Sen., D-La.), 42
Budget Committees. *See also* Congressional Budget Office (CBO)
  constraint of House Agriculture Committee by, 14
  Contract with America and, 8–9
  dominance of, 4
Budget cuts
  House Resolution (H.R.) 1976 and, 30–31
  proposed by John Kasich, 27
  proposed by Pete Domenici, 40
Budget reconciliation bill, 159
  balanced budget and, 61–62
  effect of presidential veto on, 79–80
  Freedom to Farm Act of 1995 immersion into, 66–67, 69
  House-Senate conference agreement and, 75
  relationship with commodity legislation, 81–83
  Senate approval of, 71
  Senate negotiations and, 95–97
Budget resolutions
  response to, 24–26
  role of, 23
  seven year program and, 155
Budget summit
  commodity legislation and, 81–83
  dominance of, 79–80
Bumpers, Dale (Sen., D-Ark.)
  1996 Farm Bill voting results, 121
  as chairman of the Senate Committee on Appropriations, 7
  government shutdown and, 90
  and Senate Resolution 1155, 42
Butz, Earl, 8–9

Canola. *See* Oilseed production
Chabot, Steve (Rep., R-Ohio), 113–114

Chambliss, Saxby (Rep., R-Ga.), 66
Chinese exports, 36
Clean Water Act, 144
Clinton, Bill, Pres.
  1996 Farm Bill signing by, 121
  budget reconciliation bill and, 75, 79–80, 159
  budget resolutions and, 23
  FY1996 appropriation and, 25
  government shutdown and, 89
  Senate negotiations and, 103
Closed rule, 110–112
Cloture rule, 97, 99–101
Coalition for a Competitive Food and Agriculture System. *See* Acreage Reduction Program (ARP)
Cochran, Thad (Sen., R-Miss.)
  and Agricultural Competitiveness Act (S. 1155), 57–58
  budget resolutions and, 23–24
  as chairman of the Senate Committee on Appropriations, 7
  Freedom to Farm proposal and, 56–57
  introduction of S. 1541, 97
  reduction payments and, 21
  on repeal or permanence of 1938 and 1949 acts, 118
  Senate Resolution 1155 and, 42
Colleges and universities, 17–19
Combest, Larry (Rep., R-Tex.), 66
Commission of 21st Century Production Agriculture, 56, 117
Commodity Credit Corporation (CCC), 155
  Farm Expenditures, 56
  in financing of Crop Reduction Program (CRP), 28–29
  Freedom to Farm Act of 1995 and, 34, 64
  seven year program and, 157
Commodity prices
  advantages of Freedom to Farm for, 86–87
  concerns of Democrats, 128–129
  effect of ARP and CRP on, 5–6
  effect on farm payments, 72
  estimated 1995 outlays, 56
  expiration of 1990 Farm Act and, 81
  Freedom to Farm Act of 1995 and, 64
  House-Senate conference agreement and, 75
  and House-Senate conferencing, 119
  permanent legislation and, 125
  presidential concerns of 1996 Farm Bill, 121
Commodity programs. *See also* Market analysis
  1996 Farm Act chronology, 159–160
  1996 Farm Bill and, 122
  according to USDA Blue Book, 46–47
  anonymous 1995 draft to withdrawal of, 151–154
  basic components of Freedom to Farm Act, 137–139
  Congressional Budget Office (CBO) outlay estimates for, 91–92
  Contract with America and, 8–9
  criteria for Freedom to Farm payments, 114
  criticism of FY1996 appropriation bill, 31–32
  Democratic views on, 41
  effect of budget reconciliation on, 71–72
  effect of bullish market on, 93, 105
  effect of FY 1996 appropriations on, 23
  extension of 1990 Farm Act and, 86–87
  federal outlays and, 125–126
  fixed v. variable payments, 36–37
  food assistance programs effect on 49
  Freedom to Farm Act of 1995, 34–37
  FY1996 administrative budgeting for, 29
  General Agreement on Trade and Tariffs (GATT) and, 15
  H.R. 2010 and, 55
  of John Baize, 14
  landowner/operator beneficiaries of, 13
  nonrecourse loans for, 134
  as part of balanced budget, 62
  permanent legislation of, 80–81
  possibility of Farm Acts of 1938 and 1949, 62
  powerful small groups and, 125–126
  Republican views on, 39–40
  Richard Lugar and, 9–12, 43, 141–143
  and Senate File 1155, 42
  and Senate File 1256, 41
  Senate negotiations and, 96–98
Condit, Gary A. (Rep., D-Calif.), 108, 121

180   Index

Conferencing
 description of, 114–117
 of Freedom to Farm Act, 71–73
 issues to resolve, 117
 legislative language after, 120–121
Congress. *See also* Democrats; Republicans
 Republican control of, 3–4, 6–9
 waiting due to 1995 budget summit, 79–83
Congressional Budget Office (CBO)
 1995 Senate markup voting and, 70
 "capture the baseline" and, 86–87
 Emerson-Combest proposal, H.R. 2330 and, 63, 66
 Freedom to Farm proposal and, 58–60
 Freedom to Farm proposal voting and, 66–67
 multiyear programs and, 24–26
 outlay estimation by, 90–92
 seven year baseline for CCC expenditure, 155–156
Congressional Research Service (CRS)
 extension of 1990 Farm Act and, 86–87
 Richard Lugar's questions for Senate hearings regarding, 148
Conner, Charles F., 7
Conrad, Kent (Sen., D-N.Dak.)
 1995 rebuttal to Richard Lugar, 17
 1996 Farm Bill voting results, 121
 on repeal or permanence of 1938 and 1949 acts, 118
 and Senate Resolution 1256, 41
 views on commodity programs, 40–41
Conservation
 criteria for Freedom to Farm payments, 114
 inclusion in FY1996, 42–43
 Richard Lugar's questions regarding, 144
 Senate negotiations and, 102
 Sherwood L. Boehlert's amendment and, 113
 Wetlands Reserve Program (WRP), 41
Conservation compliance. *See also* Crop insurance
 1930s dust bowl and, 135
 Freedom to Farm Act of 1995 and, 64
 Freedom to Farm Act of 1996 and, 135
 for future farm legislation, 129
 Richard Lugar's questions for Senate hearings regarding, 144
Conservation Reserve Program (CRP)
 according to S. 854, 57
 anonymous early 1995 farm program withdrawal and, 153
 capping/termination of contract, 108–109, 113, 117
 Emerson-Combest proposal, H.R. 2330 and, 63
 financing of, 28–29
 Freedom to Farm Act of 1996 and, 135
 Freedom to Farm proposal and, 57
 historical program aspects and, 135
 Large-Scale Land Idling Has Retarded Growth of U.S. Agriculture and, 4–6
 Richard Lugar's questions for Senate hearings regarding, 144
 Senate negotiations and, 99–100
Consolidated Farm Service Agency (CFSA)
 House Resolution (H.R.) 1976 and, 30
 seven year program and, 157
Contract with America, 8–9
Corn production
 and 1994-1995 commodities, 56
 and 1995 payment timing, 36
 historical program aspects and, 135
 price v. 1996 Farm Act milestones, 161
 Richard Lugar's early agenda regarding, 11–12
 Richard Lugar's questions for Senate hearings regarding, 145
 Senate negotiations and, 96
Cotton production
 and 1995 payment timing, 36
 anonymous early 1995 farm program withdrawal and, 151–152
 flex acres and, 46
 Freedom to Farm Act of 1995 and, 42, 65–66
 future farm legislation and, 131
 historical program aspects and, 135
 Joseph P. Kennedy II's amendment, 113
 marketing loans amendment for, 113–114
 nonrecourse commodity loans for, 64
 Richard Lugar's early agenda regarding, 10–12
 Richard Lugar's questions for Senate hearings regarding, 143

Senate negotiations and, 96
Coverdell, Paul (Sen., R-Ga.), 42
Craig, Larry (Sen., R-Idaho)
  introduction of S. 1541, 97
  during Senate negotiations, 98
  and Senate Resolution 1155, 42
Crop insurance
  disaster payments and, 30–31
  Freedom to Farm Act and, 138
  House Agriculture Committee markup and, 108
  House-Senate conference agreement and, 74
  Richard Lugar's questions for Senate hearings regarding, 143
  Senate negotiations and, 99
Cropland production. *See also* Conservation Reserve Program (CRP); Cropping practices
  1990 Farm Act and, 11–12
  1994/1995 commodity prices and, 56
  Large-Scale Land Idling Has Retarded Growth of U.S. Agriculture and, 3–6
  mix of total acreage base (TAB), 48
  projected (2002), 5–6
  and Richard Lugar's early agenda, 10–12
Cropping practices
  farm programs' effects on, 46
  Freedom to Farm flexibility for, 88, 94
Cuban air crisis effects, 112
Cuts in farm programs. *See also* Budget cuts
  Blue Book and, 47–49
  Kerrey-Danforth Commission report and, 16

Dairy policies
  1995 Senate markup voting and, 70
  2000 product loan program, 119–120
  anonymous early 1995 farm program withdrawal and, 152
  in Freedom to Farm proposal, 69
  future farm legislation and, 131
  House Agriculture Committee markup and, 108–109
  at House-Senate conferencing, 116
  northeast interstate dairy compact, 117
  Senate negotiations and, 97–98, 100, 102

  Solomon/Dooley proposals for, 113
  Steve Gunderson and, 55–56, 65, 107–108
Danielson, Nancy, 103
Daschle, Thomas A. (Sen., D-S.Dak.)
  1996 Farm Bill voting results, 121
  conflicting views with Patrick Leahy, 123
  legislation based on stability by, 128–129
  during Senate negotiations, 98–99, 101–102
  and Senate Resolution 1256, 41
  two-tier marketing loans and, 58, 87–88
Davis, Chester, 139. *See also* New Deal
de la Garza, E. (Rep., D-Tex.)
  1996 Farm Bill voting results, 121
  as chairman of House Agriculture Committee, 7
  investments amendment, 113
Deficiency payments
  1990 Farm Act and, 11–12
  anonymous early 1995 farm program withdrawal and, 151–153
  effect of target price phaseback on, 62
  Emerson-Combest proposal, H.R. 2330 and, 63
  Freedom to Farm proposal and, 57
  FY1996 appropriations bill and, 33
  House-Senate conferencing effect on, 75
  Price and Income Stability Working Group and, 18–19
  S. 1155 and S. 1256 and, 59
  Senate Resolution 1155 and, 42
  seven year program and, 157
  two-tier marketing loans and, 58, 87–88
  for wheat, 41
DeLay, Tom (Rep., R-Tex.), 67–68
Democrats. *See also* Voting results
  1996 Farm Bill voting results, 121
  bipartisan farm program views by, 137
  budget reconciliation Senate negotiations and, 95–97
  concerns of price variability, 128
  delaying of vote by, 109
  House Agriculture Committee markup and, 108–109
  House negotiations and, 106–107
  inability to formulate farm act pro-

Democrats (*Continued*)
  posal, 65–66
  two-tier marketing loans and, 88
  views on commodity programs, 40–41
  voting records in Fall 1995, 66–67
Disaster Feed Assistance, 30
Disaster payments. *See also* Crop insurance
  Freedom to Farm Act and, 138
  House Resolution (H.R.) 1976 and, 30–31
  Richard Lugar's questions for Senate hearings regarding, 143
Dole, Robert (Sen., R-Kans.)
  introduction of S. 1541, 97
  Senate negotiations and, 95–101, 103
Domenici, Pete V. (Sen., R-N.Mex.)
  budget cuts and, 22–23, 40
  on fixed payment approach, 89
  and Richard Lugar's early agenda, 10
Dooley, Calvin M. (Rep., D-Calif.)
  dairy policies and, 113
  House Agriculture Committee markup and, 108
Dorgan, Byron (Sen., D-N.Dak.)
  1996 Farm Bill voting results, 121
  introduction of S.1523, 97
  during Senate negotiations, 98, 100
  and Senate Resolution 1256, 41
  views on commodity programs, 41
Durbin, Richard J. (Rep., D-Ill.)
  as chairman of the House Committee on Appropriations, 7
  and House Resolution (H.R.) 1976, 30–31
Dust bowl conservation programs, 135

Economics. *See* Agricultural economics
Eisenhower soil bank, 135
Electronic Benefit Transfer, 147
Emergency livestock feed program
  and 1938 and 1949 acts, 119
  1995 Senate markup voting and, 70
  Emerson-Combest proposal, H.R. 2330 and, 63
Emerson, Bill (Rep., R-Mo.), 66
Emerson-Combest proposal, H.R. 2330, 63, 66
Entitlement programs, 21–26
Environmental concerns

Conservation Reserve Program (CRP) and, 57
  for future farm legislation, 129
  House negotiations and, 106
  objections to H.R. 2854 regarding, 113–114
  of outlay estimations, 92
  Richard Lugar's questions for Senate hearings regarding, 144
  Senate negotiations and, 97–98
  seven year program and, 156
Environmental Quality Incentive Program (EQUIP)
  1995 Senate markup voting and, 70
  Senate negotiations and, 99
  Sherwood L. Boehlert's amendment, 113
  size of livestock operations and, 117
Espy, Mike, 15
Ethanol production, 137
Exon, James (Sen., D-Nebr.)
  1996 Farm Bill voting results, 121
  and Senate Resolution 1256, 41
Expanded acreage options, 5–6
Export Enhancement Program (EEP)
  1995 Senate markup voting and, 70
  Emerson-Combest proposal, H.R. 2330 and, 63
  FY1996 budgeting and, 30
  House-Senate conferencing effect on, 73
  Richard Lugar's questions for Senate hearings regarding, 144–146
Exports
  Freedom to Farm Act of 1995 and, 36
  subsidy programs (1994) for, 15
Extension Service, 141

Family farms
  Richard Lugar's questions for Senate hearings regarding, 144–146
  targeting payments to, 17
  two-tier marketing loans and, 88
  wheat deficiency payments for, 41
Farm Act of 1938, 62, 80–81, 125
Farm Act of 1949, 62, 80–81, 125, 139
Farm Act of 1985, 129
Farm Act of 1990
  1995, desire to continue, 61–62, 73, 74, 88
  bullish grain markets and, 85–86

deficiency payments according to, 11–12
Freedom to Farm proposal and, 58–60
outlay estimation of, 91–92, 114
possible extension of, 86–87, 89–90, 106
Farm Act of 1996
  anonymous 1995 letters and, 3–4
  chronology, 159–160
  Freedom to Farm Act of 1995 and, 76–77
Farm Act of 2002
  budget resolutions and, 127
  economic health of agriculture and, 127–129
  factors effecting, 125–127
Farm Bill 1995: Guidance of the Administration, 46–47
Farm Bureau. *See* American Farm Bureau Federation (AFBF)
Farm Credit System, 148
Farmer Mac, 150
Farmer-owned reserve, 138–139, 152
Farmers Home Administration (FmHA), 141, 147–149
Farmers Union, 103
Farm operators. *See* Operators
Farm owners. *See* Landowners
Federal Agricultural Mortgage Corporation, 149
Federal Agriculture Improvement and Reform Act of 1996. *See* Farm Act of 1996
Federal Crop Insurance Fund (FCIF), 29, 141. *See also* Crop insurance
Federal Grain Inspection, 141
Federal transfers, 92–93
Feed grain production
  anonymous early 1995 farm program withdrawal and, 151–152
  farmer-owned reserve and, 138–139
  flex acres and, 46
  historical program aspects and, 135
  nonrecourse commodity loans for, 64
  and Sen. Lugar's early agenda, 10–12
  and USDA publications, 18
Feingold, Russell D. (Sen., D-Wis.), 121
Flex acres. *See also* Set-aside acres
  1995 Senate markup voting and, 70
  Agricultural Competitiveness Act (S. 1155) and, 58
  Blue Book and, 46
  Dan Glickman on, 48
  Freedom to Farm Act and, 138
  and House-Senate conferencing, 117
  Senate negotiations and, 99
Florida Everglades purchase, 113
Food and Drug Administration (FDA), 141
Food and Nutrition Service, 141
Food assistance programs
  balanced budget and, 49–50
  bipartisan farm program views and, 137
  blocking of, 50–53
  comparison with farm programs, 8–9
  dairy policies and, 108
  effect on farm commodity programs, 49
  Freedom to Farm proposal and, 57
  FY1996 budgeting for, 29
  and House-Senate conferencing, 117
  inclusion in FY1996, 42–43
  international, 113
  Republican's proposed budget levels for, 45
  Richard Lugar's questions for Senate hearings regarding, 146–147
  Senate negotiations and, 99–100
Food Safety and Inspection Service, 141
Food stamps. *See* Food assistance programs
Ford, Wendell H. (Sen., D-Ky.), 41
Foreign Agriculture Service, 141
Forest Service, 141
Frank, Barney (Rep., D-Mass.), 113
Freedom to Farm Act of 1995
  anonymous summer draft, 155–157
  dairy policy in, 65
  description, 34–37, 64–65
  House-Senate conferencing, 71–73
  voting on Democrats' proposal, 66
  voting on Emerson-Combest bill, 66
Future farm legislation. *See also* Farm Act of 2002
  commodity programs and, 125–126
  impact of political power on, 129–130
  industrialized agriculture and, 132
  operator-owned land and, 130–131

Gardner, Bruce, 73
General Agreement on Trade and Tariffs (GATT), 15

184 Index

Gingrich, Newt (Rep., R-Ga.)
 Freedom to Farm proposal and, 67–68, 69
 prediction of bill passage, 109
 support of Florida Everglades property by, 113
Glenn, John (Sen., D-Ohio), 121
Glickman, Dan, 45
 on base acreage increase, 48
 concerns of 1996 Farm Bill, 121
 on extension of 1990 Farm Act, 86–87
 Senate negotiations and, 100
Government intervention
 1996 shutdown, 89
 anonymous 1995 draft to withdrawal of, 151–154
 anonymous 1995 letters and, 3–4, 13–14
 Contract with America and, 8
Grains. *See* Feed grain production
Gramm, Phil (Sen., R-Tex.), 101
Grassley, Charles (Sen., R-Iowa), 41–42
Great Plains Conservation Program, 30
GSM-102 program, 146
Gunderson, Steve (Rep., R-Wis.)
 dairy policy proposal by, 55–56, 107–108
 dairy price support elimination and, 65
 Solomon/Dooley dairy bill and, 112–113

Harkin, Tom (Sen., D-Iowa)
 1996 Farm Bill voting results, 121
 on repeal or permanence of 1938 and 1949 acts, 118
 and Senate Resolution 1256, 41
Hatch, Orrin G. (Sen., R-Utah), 101–102
Hatfield, Mark (Sen., R-Oreg.), 24
Helms, Jesse (Sen., R-N.C.), 42
Historical aspects
 agenda of Sen. Richard G. Lugar, (R-Kans.), 3–4
 Agricultural Adjustment Act of 1938, 134
 anonymous 1995 letter and, 3–4, 13–14
 bipartisan farm program views, 137
 constraints by Budget Committees, 14
 economic collapse of 1920s, 134
 effects of WWI and II, 139

 release of *Large-Scale Land Idling Has Retarded Growth of U.S. Agriculture*, 3–6
 Republican control of Congress, 3–4, 7–9
 WWII war economy and, 135–136
Honey production
 FY1994/95 appropriations for, 28
 historical program aspects and, 135
House Agriculture Committee
 Bill Barrett and, 36
 Freedom to Farm proposal and, 57
 House negotiations and, 106
 House-Senate conference agreement of, 74–75
 markup attempt by, 65–69, 108–109
 primacy establishment of, 27–28
 release of supporters of Freedom to Farm by, 92–93
 Republican control of Congress and, 3–4
House Budget Committee, 39–40
House Committee on Agriculture, 65
House of Representatives
 1996 voting results, 121
 Democrats proposal for 1995, 63
 Emerson-Combest proposal, H.R. 2330 for 1995, 63
 food program blocking and, 52–53
 House Committee on Appropriations, 7
 H.R. 2854 and, 110
 markup attempt by House Agriculture Committee, 65–69, 108–109
 readying for Freedom to Farm debates, 86–88
 reason for Freedom to Farm support, 67–68
House Resolution (H.R.) 1976
 amendments to, 31–33
 description, 30
House Resolution (H.R.) 2010, 55. *See also* Freedom to Farm Act of 1995
House Resolution (H.R.) 2147, 57
House Resolution (H.R.) 2195. *See* Freedom to Farm Act of 1995
House Resolution (H.R.) 2330. *See* Emerson-Combest proposal
House Resolution (H.R.) 2854
 House of Representatives' voting on, 110, 112–114
 Pat Roberts and, 105

House Rules Committee amendment selection, 110–112
House-Senate conferencing. *See* Conferencing
Housing programs
  FY1996 appropriations bill and, 32–33
  House Resolution (H.R.) 1976 and, 30–31
Hubert H. Humphrey Institute of Public Affairs, 18–19

Idling of acres, 35–37. *See also* Acreage Reduction Program (ARP); Conservation Reserve Program CRP)
Inouye, Daniel K. (Sen., D-Hawaii), 42
Insurance industry
  Emerson-Combest proposal, H.R. 2330 and, 63
  House-Senate conference agreement and, 74
  H.R. 2147 and, 57
International concerns
  1995 statement by Kent Conrad, 17
  Chinese exports, 36
  and effects on Farm Act of 2002, 126
  price enhancement, 136–137
  Richard Lugar's questions for Senate hearings regarding, 144–146

Johnston, J. Bennett (Sen., D-La), 42

Kasich, John R. (Rep., R-Ohio)
  1995 budget cuts proposed by, 27
  budget cuts and, 22–23
  reporting of law changes for outlays and, 67
Kennedy, Joseph P. II. (Rep., D-Mass.), 112
Kerrey, J. Robert (Sen., D-Nebr.)
  1996 Farm Bill voting results, 121
  and Senate Resolution 1256, 41
Kerrey-Danforth Commission report, 16
Kohl, Herb (Sen., D-Wis.), 121

Land grant colleges, 17–19
Landowners
  as beneficiaries of commodity programs, 13–14
  benefits of Freedom to Farm for, 85–86, 93
  Blue Book and, 47
  future farm legislation and, 126, 130–131
  House-Senate conferencing effect on, 73, 75
  maximum payments to, 90
  proposed stability of Freedom to Farm proposal, 68
Land value
  commodity program benefits to, 13–14
  effect of target price phaseback on, 62
  Freedom to Farm Act of 1995 and, 37
  Richard Lugar's questions for Senate hearings regarding, 142
Large-Scale Land Idling Has Retarded Growth of U.S. Agriculture, 3–6, 159
Leahy, Patrick J. (Sen., D-Vt.), 121, 159–160
  1994 Senate letter by, 19–20
  1995 view on commodity program, 43
  budget cuts and, 48
  as chairman of Senate Agriculture Committee, 7
  conflicting views with Thomas Daschle, 123
  flexibility/guaranteed payments contributions of, 95–97
  at House-Senate conferencing, 116
  during Senate negotiations, 98–99, 101–104
  views on commodity programs, 42
Leesburg, Va., 1995 conference, 27
Livestock operations
  Environmental Quality Incentive Program (EQUIP) and, 117
  historical program aspects and, 135
Livingston, Bob (Rep., R-La)
  concerns of 1996 Farm Bill by, 122
  objections to conservation amendment, 113
  thoughts against farm bill by, 110
Loan rates
  anonymous early 1995 farm program withdrawal and, 151
  historical program aspects and, 135
  House-Senate conference agreement and, 74
  nonrecourse, 134

186   Index

Loan rates (*Continued*)
  Richard Lugar's questions for Senate hearings regarding, 146
Lott, Trent (Sen., R-Miss.)
  Senate negotiations and, 101
  and Senate Resolution 1155, 42
Lowey, Nita M. (Rep., D-N.Y.)
  as critic of FY1996 appropriations bill, 31–33
  peanut program by, 113
Lugar, Richard G. (Sen., R-Ind.)
  1994 Senate letter by, 19–20
  1995 proposed outlay reduction by, 16–17
  1995 view on commodity program, 39–40, 43
  1996 Farm Bill conservation concerns by, 122
  Blue Book and, 47
  as chairman of Senate Agriculture Committee, 3, 7, 159–160
  commodity program questions by, 141–143
  conservation program questions by, 144
  early agenda of, 3–4, 9–12, 21
  export program questions by, 144–146
  food program blocking and, 51–52
  at House-Senate conferencing, 116
  introduction of S. 1541, 97
  nutrition program questions by, 146–147
  phasing out of commodity programs by, 131–132
  on repeal or permanence of 1938 and 1949 acts, 118
  rural development program questions by, 147–149
  Senate negotiations and, 95–96, 100–102

Mack, Connie (Sen., R-Fla.), 42
Market analysis
  Congressional Budget Office (CBO) and, 25
  dairy production separation, 65
  House of Representatives' support of proposal and, 68
  no government intervention and, 151
  payment timing and, 36–37

Market Promotion Program (MPP), 33, 70
Market transition contracts
  Freedom to Farm Act of 1995 and, 64
  House-Senate conference agreement and, 74–75
Marketing loans (two-tier), 58, 87–88
  Freedom to Farm Act and, 139
  Richard Lugar's questions for Senate hearings regarding, 143
Markup attempts
  by House Agriculture Committee, 65–69
  by Pat Roberts, 108–109
McCain, John (Sen., R-Ariz.), 121
McConnell, Mitch (Sen., R-Ky.), 42
Medicare/Medicaid
  effect on Farm Act of 2002, 127
  House blocking of, 52
Milk production. *See also* Dairy policies
  historical program aspects and, 135
  Richard Lugar's questions for Senate hearings regarding, 143
  Steve Gunderson and, 55–56, 65, 107–108, 113
Miller, Dan (Rep., R-Fla.), 31–33
Mink production, 33
Mitchell, Gary, 7–8
Multiyear farm programs' outlay estimate, 22, 24–25

National Center for Food and Agricultural Policy, 18–19
National Corn Growers Association, 92
National Grain and Feed Association, 4–6, 159
Natural Resources Conservation Service, 24–25, 155
New Deal, 137
No policy change option, 5–6
Nonrecourse loans, 134
  1950s grain accumulation due to, 136
  anonymous early 1995 farm program withdrawal and, 151, 153
  farmer-owned reserve and, 138–139
  historical program aspects and, 135
North Central Regional Research Project, 17–19
Nunn, Sam (Sen., D-Ga.), 42

Off-farm income, payment ban, 48, 87
Office of Management and Budget (OMB)
  Alice Rivlin of, 15
  during Senate negotiations, 98
Oilseed production
  anonymous early 1995 farm program withdrawal and, 151–152
  nonrecourse commodity loans for, 64
  and Richard Lugar's early agenda, 10–12
  Richard Lugar's questions for Senate hearings regarding, 145
Operators
  anonymous early 1995 farm program withdrawal and, 151
  as beneficiaries of commodity programs, 13–14
  benefits of Freedom to Farm for, 85–86, 93
  elimination of supply management and, 156
  future farm legislation and, 126, 130
  House-Senate conferencing effect on, 73, 75
  maximum payments to, 90
  off-farm income and, 48, 87
  payment based on production volume for, 123
  program flexibility for, 58–60
  proposed stability of Freedom to Farm proposal, 68
  Richard Lugar's questions for Senate hearings regarding, 142
Outlays. *See* Congressional Budget Office (CBO)
  in 1996 Farm Bill, 122
  farm commodity programs and, 125–126
  result of Republicans decrease of, 114

Packers and Stockyards Administration, 141
Packwood, Robert (Sen., R-Oreg.), 51–52
Panetta, Leon (chief of staff), 88
Payment yield. *See also* Commodity prices
  of 1990 Farm Act, 11–12
  according to Blue Book, 46
  based on production volume, 123
  bullish grain markets and, 85–86
  debate of timing of, 36–37
  effect of $5 billion cut on, 87
  for family farms, 17, 41
  Freedom to Farm Act and, 34, 64, 136, 138
  Senate negotiations and, 99–100, 103
  seven year program and, 156
Peanut production
  Agricultural Competitiveness Act (S. 1155) and, 58
  anonymous early 1995 farm program withdrawal and, 152–153
  budget outlay changes and, 87
  complexity of program, 120
  Emerson-Combest proposal, H.R. 2330 and, 63
  FY1996 appropriations bill and, 31
  historical program aspects and, 135
  House Agriculture Committee markup and, 108
  H.R. 2854 and, 112, 113
  Richard Lugar's questions for Senate hearings regarding, 143
  Rick Santorum's markup voting and, 70–71
  Senate negotiations and, 102
Peterson, Collin C. (Rep., D-Minn.), 108
P.L. 480 Title I program, 146
President Bill Clinton. *See* Clinton, Bill, Pres.
Price and Income Stability Working Group, 18–19
Prices. *See also* Commodity prices; Payment yield
  1990 Farm Act and, 11–12
  anonymous early 1995 farm program withdrawal and, 151
Price supports
  anonymous early 1995 farm program withdrawal and, 151, 153
  Price and Income Stability Working Group and, 18–19
Production. *See specific commodities, e.g.*, Cotton production; Cropland production
Pryor, David (Sen., D-Ark.)
  1996 Farm Bill voting results, 121
  and Agricultural Competitiveness Act (S. 1155), 57–58

Pryor, David (Sen., D-Ark.) (*Continued*)
  government shutdown and, 90
  and Senate Resolution 1155, 42

Reconciliation bill. *See* Budget reconciliation bill
Reduction payments, 21
Renters. *See* Operators
Representatives and senators, xiii–xv. *See also names of individuals*
Republicans. *See also* Voting results
  1995 delay of appropriation bills by, 80
  1996 Farm Bill voting results, 121
  bipartisan farm program views by, 137
  Contract with America and, 8–9
  control of Congress, 22
  decrease of outlays by, 114
  food program blocking and, 51–53
  government shutdown and, 89
  markup attempt by House Agriculture Committee, 68–69
  threatened USDA shutdown by, 80
  views on commodity programs, 39–40
Research
  House Agriculture Committee markup and, 108
  inclusion in FY1996, 42–43
  money diversion to, 90
  Senate negotiations and, 102
Rice production
  and 1994-1995 commodities, 56
  anonymous early 1995 farm program withdrawal and, 151–152
  flex acres and, 46
  Freedom to Farm Act of 1995 and, 42, 65–66
  historical program aspects and, 135
  and House-Senate conferencing, 119
  nonrecourse commodity loans for, 64
  and Richard Lugar's early agenda, 10–12
  Richard Lugar's questions for Senate hearings regarding, 143
  Senate negotiations and, 96, 101–102
Risk management, 141–142
Rivlin, Alice, 15
Roberts, Pat (Rep., R-Kans.)
  as chairman of House Agriculture Committee, 3, 7, 159–160
  early agenda for Farm Bill debate, 21
  on fixed payment approach, 89
  food program blocking and, 52
  Freedom to Farm Act of 1995 and, 35–37
  Freedom to Farm proposal and, 13–14, 56–57
  FY1996 appropriations debate and, 26–27
  House Agriculture Committee markup and, 108–109
  House negotiations and, 105–106
  House-Senate conferencing and, 74–77, 116
  H.R. 2147 and, 57
  H.R. 2854 and, 105, 113–114
  moving to the Senate, 130
  on repeal or permanence of 1938 and 1949 acts, 118
  Senate negotiations and, 96–97
Rural development programs
  FY1996 appropriations bill and, 32
  House Agriculture Committee markup and, 108
  House negotiations and, 106
  and House-Senate conferencing, 117
  money diversion to, 90
  outlay estimation concerns of, 92
  Richard Lugar's questions for Senate hearings regarding, 147–149
  Senate negotiations and, 102
  USDA in 2002 and, 126
Rural Electrification Administration (REA), 141, 147–149

Santorum, Rick (Sen., R-Pa.)
  1995 Senate markup voting by, 70–71
  Senate negotiations and, 102
  views on commodity programs, 40
School lunch programs, 52. *See also* Food assistance programs
Schumer, Charles E. (Rep., D-N.Y.), 113–114
Secretary of Agriculture. *See* Espy, Mike; Glickman, Dan
Senate
  1996 voting results, 121
  chronicle of Senate negotiations, 97–104
  food program blocking and, 51–52

Index 189

markup attempt by Senate Agriculture Committee, 70–71
readying for Freedom to Farm debates, 86–88
Senate Agriculture Committee
1994 Leahy/Lugar letter and, 19–20
Democratic commodity views in, 40–41
House-Senate conference agreement of, 74–75
markup attempt by, 70–71
Republican commodity views in, 39–40
Republican control of Congress and, 3
Senate Budget Committee, 22–23, 40
Senate Resolution 1155, 42
description, 57–58
Senate Resolution 1256, 41
Senate Resolution 1541, 101
Senate Resolution 854, 57
Senators and representatives, xiii–xv. *See also names of individuals*
Set-aside acres, 155
1990 Farm Act and, 11–12
Freedom to Farm Act of 1995 and, 64–65. *See also* Acreage Reduction Program (ARP)
Seven Year Contract. *See also* Freedom to Farm Act of 1995
advantages, 156–157
disadvantages, 157
Shays, Christopher (Rep., R-Conn.), 113
Skeen, Joe (Rep., R-N.Mex.), 159
budget resolutions and, 23–24
as chairman of the House Committee on Appropriations, 7
concerns of 1996 Farm Bill by, 110, 122
FY1996 appropriations debate and, 26–27
Social Security effect, 127
Solomon, Gerald (Rep., R-N.Y.), 69, 109–110, 112–113
Soybean production. *See also* Oilseed production
and 1994-1995 commodities, 56
loan rate of S. 1155 and, 58
Sparks Commodities, 101
Special Supplemental Nutrition Program for Women, Infants, and Children (WIC)
dairy policies and, 108

FY1996 appropriations bill and, 32–33
FY1996 budgeting for, 29
Richard Lugar's questions for Senate hearings regarding, 147
Staff budgeting, 29
Stenholm, Charles W. (Rep., D-Tex.)
1996 Farm Bill voting results, 121
on budget balancing, 87
House negotiations and, 107
Subcommittees
FY1996 appropriation, 25–26
on General Farm Commodities, 36
on Livestock, Dairy and Poultry, 65
rival with Agriculture Committee, 27–31
Sugar production
Agricultural Competitiveness Act (S. 1155) and, 58
amendments for, 114
anonymous early 1995 farm program withdrawal and, 152–153
Cuban air crisis and, 112
FY1996 appropriations bill and, 31
historical program aspects and, 135
H.R. 2854 and, 112
purchase of Florida Everglades and, 113
Richard Lugar's questions for Senate hearings regarding, 143
Rick Santorum's markup voting and, 70–71
Sunset provisions, 125
Suspension of Certain Provisions Regarding Program Crops, 56
Swampbuster provision, 144

Talmadge, Herman, 115
Targeted market loans, 58
Target prices
Agricultural Competitiveness Act (S. 1155) and, 58
anonymous early 1995 farm program withdrawal and, 152
House-Senate conference agreement and, 74–75
Kerrey-Danforth Commission report and, 16
reduction of, 11–12, 21
Richard Lugar and, 21, 47, 62, 159
two-tier marketing loans and, 58, 88

Telephone program (rural), 148
Thurmond, Strom (Sen., R-S.C.), 42
Title I program (P.L. 480), 146
Tobacco production
   anonymous early 1995 farm program withdrawal and, 152–153
   FY1996 appropriations bill and, 32–33
   historical program aspects and, 135
   program termination and, 87
   Richard Lugar's questions for Senate hearings regarding, 143
Tolley, Howard, 139–140. *See also* New Deal
Total acreage base (TAB)
   Blue Book and, 46
   mix of, 48
Trade agreements/concerns. *See* International concerns
Two-tier marketing loans. *See* Marketing loans (two-tier)

United States Department of Agriculture (USDA)
   2002, rural development programs in, 126
   Blue Book of, 45–47
   and budget assignments to Agricultural Appropriations Subcommittee, 29
   budget resolutions and, 24–25
   estimated 1995 outlays, 56
   FY1996 appropriations bill and, 32–33
   initiation of year 2000 dairy product loan program, 119–120
   publications regarding farm bill, 18
   research money from, 88
   Richard Lugar's questions for Senate hearings regarding, 145
   threatened shutdown of, 79–80
Uruguay Round, 145

Veto of reconciliation bill, 79–80
Voting results
   House Agriculture Committee markup, 108–109
   House of Representatives' 1995 results, 63–68
   House of Representatives' 1996 results, 105
   at House-Senate conferencing chambers, 121
   on H.R. 2854, 112–114
   on repeal or permanence of 1938 and 1949 acts, 118–119
   Senate's 1995 results, 70–71

Wallace, Henry, 128, 134
Warner, John W. (Sen., R-Va.), 42
Welfare. *See also* Food assistance programs
   comparison with landowners, 47
   criticism of Freedom to Farm proposal as, 68
   Freedom to Farm Act of 1995 and, 34–35
   seven year program and, 157
Wellstone, Paul D. (Sen., D-Minn.), 41
Wetlands Reserve Program (WRP), 29–30
Wheat production
   and 1994-1995 commodities, 56
   and 1995 payment timing, 36
   anonymous early 1995 farm program withdrawal and, 151–152
   flex acres and, 46
   historical program aspects and, 135
   and Kent Conrad, 40–41
   nonrecourse commodity loans for, 64
   and Richard Lugar's early agenda, 10–12
   Richard Lugar's questions for Senate hearings regarding, 144–145
   Senate negotiations and, 96
   and USDA publications, 18
Whitten, Jamie, (Rep., D-Miss), 28
Whole farm base, 157
Wool production, 135

Zero ARP. *See* Expanded acreage options
Zimmer, Dick (Rep., R-N.J.), 55

ISBN 0-8138-2608-X

343.73 Sch
Schertz, Lyle P.
The making of the 1996 Farm Act

DATE DUE

PROPERTY OF
CENTRAL COMMUNITY COLLEGE
HASTINGS, CAMPUS

WITHDRAWN